Computational Approaches to Cognition and Perception

Editor-in-chief
Amy H. Criss, Department of Psychology, Syracuse University, Syracuse, New York, USA

Computational Approaches to Cognition and Perception is a series that aims to publish books that represent comprehensive, up-to-date overviews of specific research and developments as it applies to cognitive and theoretical psychology. The series as a whole provides a rich foundation, with an emphasis on computational methods and their application to various fields of psychology. Works exploring decision-making, problem solving, learning, memory, and language are of particular interest. Submitted works will be considered as well as solicited manuscripts, with all be subject to external peer review.

Books in this series serve as must-have resources for Upper-level undergraduate and graduate students of cognitive psychology, theoretical psychology, and mathematical psychology. Books in this series will also be useful supplementary material for doctoral students and post-docs, and researchers in academic settings.

More information about this series at http://www.springer.com/series/15340

James J. Palestro • Per B. Sederberg
Adam F. Osth • Trisha Van Zandt
Brandon M. Turner

Likelihood-Free Methods for Cognitive Science

James J. Palestro
Department of Psychology
The Ohio State University
Columbus, OH, USA

Per B. Sederberg
Department of Psychology
The Ohio State University
Columbus, OH, USA

Adam F. Osth
University of Melbourne
Parkville, VIC, Australia

Trisha Van Zandt
Department of Psychology
The Ohio State University
Columbus, OH, USA

Brandon M. Turner
Department of Psychology
The Ohio State University
Columbus, OH, USA

ISSN 2510-1889   ISSN 2510-1897 (electronic)
Computational Approaches to Cognition and Perception
ISBN 978-3-319-72424-9   ISBN 978-3-319-72425-6 (eBook)
https://doi.org/10.1007/978-3-319-72425-6

Library of Congress Control Number: 2017962914

Printed on acid-free paper

This Springer imprint is published by Springer Nature
The registered company is Springer International Publishing AG
The registered company address is: Gewerbestrasse 11, 6330 Cham, Switzerland

Foreword

This book provides a concise overview of recent developments in likelihood-free inference, thereby opening a new chapter in the field of cognitive modeling. With the easy availability of computers, researchers in the field introduced a glut of mechanistic models of cognition that have no closed-form expression of the likelihood function, placing them outside of the standard statistical realm. As such, it is not generally possible to fit a mechanistic model to observed data by maximum likelihood estimation or Markov chain Monte Carlo based Bayesian methods.

Instead, as a provisional step, the best-fitting parameter values of the model of interest were estimated in the frequentist framework by a brute-force, computationally intensive implementation of least squares estimation. Obtaining the sampling distribution of a test statistic, however, is out of question. The lack of a test statistic hinders the progress of model development because without it one can neither properly assess the adequacy of the model nor compare the fitted model with other competing ones. Things are much the same in the Bayesian framework. Consequently, the technical challenge of likelihood-free inference has been a stubborn barrier to making theoretical progress in the field. This unhappy state of affairs has changed dramatically in the early 2000s with the introduction of a series of likelihood-free algorithms, such as approximate Bayesian computation (ABC), Gibbs ABC, synthetic likelihood, and probability density approximation. With the availability of these easy-to-implement methods, model evaluation and comparison for mechanistic models are now well within the reach of every modeling scientist.

The senior author of this book has been at the forefront in the development and application of likelihood-free methods to analyzing and fitting psychological models, initially applied to models of recognition memory as part of his dissertation research at the Ohio State University and subsequently expanded to other domains of modeling including episodic memory and perceptual decision making.

The book itself, written based on the authors' published articles, offers a technically comprehensive yet clearly explained account of all major likelihood-free algorithms, with a focus on Bayesian inference. Another noteworthy feature of the book is its practically oriented approach in which application examples of the algorithms for well-known models in psychology are discussed in great detail with accompanying pseudocode and actual code as well. The work such as this will make a transformative impact on the advancement and practice of computational modeling of cognition. This book is a first positive step in that direction.

The Ohio State University   Jay Myung
Columbus, OH, USA
September 2017

Acknowledgments

This work was supported by an NSF grant SES-1424481 awarded to Van Zandt, and an NRSA grant 1F32GM103288-01 awarded to Turner.

Contents

About the Authors

James J. Palestro is a doctoral student in the Psychology Department at the Ohio State University. He received a B.A. from Youngstown State University in psychology in 2012 and an M.A. from the Ohio State University in 2018. His research interests include cognitive modeling and the neural bases of perceptual decision making.

Per B. Sederberg is an Associate Professor in the Department of Psychology at the University of Virginia. He received his undergraduate degree in cognitive science from the University of Virginia in 1996. He then worked as a computer programmer in industry before returning to earn his Ph.D. in neuroscience from the University of Pennsylvania in 2006. He spent 4 years as a postdoctoral fellow at Princeton University where he studied machine-learning applications to the study of the brain. In 2010 he began as an Assistant Professor at the Ohio State University and moved to the University of Virginia in the summer of 2017. As a computational cognitive neuroscientist, his research seeks to develop a comprehensive mathematical understanding of cognition, with a particular focus on human memory and decision making, that links neuroscience and behavior.

Adam F. Osth is a Lecturer in the Melbourne School of Psychological Sciences at the University of Melbourne. He received a B.A. in psychology from the University of California Santa Cruz, and both an M.A. in 2011 and Ph.D. in 2014 in psychology from the Ohio State University. He spent 2 years as a postdoctoral fellow at the University of Newcastle in New South Wales, Australia. His research focuses on both empirical and computational modeling investigations of episodic memory, with focuses in recognition memory, serial order memory, and the integration of models of memory and decision making.

Trisha Van Zandt is a Professor of Psychology at the Ohio State University. She is a member of the Society for Mathematical Psychology, of which she was President in 2006–2007, and the American Statistical Association. She has received multiple research grants from the National Science Foundation and the Presidential

Early Career Award for Scientists and Engineers in 1997. She is coauthor of review chapters "Designs for and Analyses of Response Time Experiments" in the Oxford Handbook of Quantitative Methods and "Mathematical Psychology" in the APA Handbook of Research Methods in Psychology.

Brandon M. Turner is an Assistant Professor in the Psychology Department at the Ohio State University. He received a B.S. from Missouri State University in mathematics and psychology in 2008, an MAS from the Ohio State University in statistics in 2010, and a Ph.D. from the Ohio State University in 2011. He then spent 1 year as a postdoctoral researcher at the University of California, Irvine, and 2 years as a postdoctoral fellow at Stanford University. His research interests include dynamic models of cognition and perceptual decision making, efficient methods for performing likelihood-free and likelihood-informed Bayesian inference, and unifying behavioral and neural explanations of cognition.

Motivation

What I cannot create, I do not understand.

Richard Feynman

The ultimate endeavor of any good cognitive scientist is to build a model that mimics the essential dynamics of the (human) mind. We should not hope to create a model that is perfectly accurate, as the only model that could perfectly reproduce any dynamic is one that is as complex as the original system; a philosophical argument known as Bonini's paradox. We seek out and develop models because they are our best hope for generalizing complex decision making processes across individuals, tasks, or time. Our criterion for evaluating the utility of a model is not only in what it provides in terms of understanding, but also in how well it can capture essential trends in behavioral data.

A psychological model is a representation of a psychological process, a representation that quantifies or provides a mechanism for how a behavioral task is performed. When we write down a model, we write down a statement of the form

$$y = f(x, \theta).$$

where y is a set of dependent variables or observed measurements, x is a set of independent variables or experimental manipulations, the function f is the mathematical structure dictated by our theory of how the data y are generated, and θ is a set of parameters that relate the independent variables to the model structures. When we want to explore how well the model explains our data y, or how well it predicts new data y', we fit the model to the data y by estimating the parameters θ.

These two endeavors—explanation and prediction—are often considered the foundational pillars of cognitive modeling [1]. Both endeavors are facilitated by accurate cognitive models, and both require detailed knowledge of estimated parameters. Hence, one of the major priorities of any would-be cognitive modeler

© Springer International Publishing AG 2018
J.J. Palestro et al., *Likelihood-Free Methods for Cognitive Science*,
Computational Approaches to Cognition and Perception,
https://doi.org/10.1007/978-3-319-72425-6_1

should be accurate parameter estimation—the method for finding the parameter value or values that best fit the observed data.

Consider a recognition task, in which people are asked to discriminate between items that were presented to them earlier in an experiment (targets) from items that were not (distractors). An old, well-known model of how recognition might be accomplished is the "high threshold" model [2, 3]. This model is based on the idea that if an item were presented to a person earlier in the experiment, it will have left behind a memory trace. What is the function f, or model structure that relates our experimental manipulations x and parameters θ to the dependent variable y, which, in this case, is the probability of identifying an item as "old"?

The high-threshold model has two parameters. The first, R, is the probability that an old target item leaves behind a trace. If a trace is present, then a person will always respond "old." The second, g, is the probability that a person responds "old" by guessing alone. If a trace is not present, then the person will respond old with probability g. Therefore, the probability of responding "old" to a target item is

$$R + g(1 - R).$$

When an item is new (i.e., a distractor), there is no trace, and the process reduces to guessing alone. So the probability of responding "old" to a distractor item is g. The function relating these parameters to recognition performance is therefore

$$f(x, \{R, g\}) = \begin{cases} R + g(1 - R) & \text{if } x \text{ is a target} \\ g & \text{if } x \text{ is a distractor,} \end{cases}$$

where x plays the role of an independent variable that identifies the item as a target or distractor.

The parameters R and g, together with the probability structure $f(x, \{R, g\})$, provide the explanatory mechanisms of the high-threshold model. Knowing how small or large R is tells us something about the efficiency of the memory process. Knowing how small or large g is tells us the tendency of the person to respond "old" or "new." Furthermore, we might expect R and g to be different over different individuals. Some people have better memories than others; some people are less willing to say "old" when they aren't certain. Information about these parameters, how they change over different experimental conditions, and how they differ over individuals, is critical to being able to test the model, decide if it is a good model or not, and in so doing learn about the psychological process that generated the data.

This brings us to the problem that is the focus of this book: how do we learn about the parameters of the model? The process of model fitting can be conducted in a number of ways, all of them "correct" in some sense, but some better than others in the context of different modeling goals. We can divide these methods into two general types: frequentist and Bayesian. The reader is probably already familiar with this distinction, but it is important to highlight the key differences between the two approaches.

Frequentist inference treats data as random and parameters as fixed quantities to be estimated. Parameters are assumed to be fixed within a group, condition, or block of experimental trials, and inference is therefore based on the sample space of hypothetical outcomes that might be observed by replicating the experiment many times. Inference about these unknown, fixed parameters takes the form of a null hypothesis test (such as a t-test), or estimating the parameters by determining the parameter values that minimize the difference between the model predictions and the data.

Bayesian inference treats both data and parameters as random, but after data have been obtained they are fixed. Inferences about parameters are based on the probability distributions of the parameters, distributions referred to as *posterior distributions*. It may be odd to think of a parameter as random; think, for example, about a parameter like the constant $\pi = 3.1416\ldots$, or $c = 2.99792458 \times 10^8$. Are these parameters truly random? Most people would say no. The variability in a parameter that is represented by its posterior distribution should be viewed not as true variability in the parameter's value, but as our uncertainty about its true value. But where there is uncertainty there is also information; the Bayesian approach makes use of whatever prior information is available about a parameter and incorporates that information into the inference.

Despite the very different viewpoints that frequentists and Bayesians have about parameters, their goals are the same. Both groups want to develop and test models that can explain and predict behavior. Understanding behavior means understanding how model parameters change with experimental conditions, so long as we can link those parameters to specific mechanisms.

Bayesian methods have become very popular in mathematical and computational psychology over the last decade [4–21]. The reasons for this growth in popularity are numerous, but can be linked to the wide availability of powerful computing resources and, most importantly, to the fact that Bayesian techniques work where frequentist methods cannot easily be applied. In particular, Bayesian inference can be performed in the context of models of theoretical interest, while frequentist methods often must depend on simplifying asymptotic assumptions (e.g., the central limit theorem). These models can be embedded in hierarchical structures that permit estimation of individual differences as well as overall effects of experimental manipulations, and posterior distributions permit us to examine relationships between parameters that would ordinarily be unobservable. Bayesian methods also permit comparisons across models that are very different: non-nested models, where models are more than just special cases of each other (with, say, certain parameters fixed at 0 or 1), can be compared and quantitatively evaluated; and models that differ in dimensionality, the number of unique parameters each have.

In the next sections in this chapter we will present the most common methods for parameter estimation and contrast them with Bayesian methods. We will then discuss the problem that is the central focus of this book: how do we estimate the parameters of a model whose predictions can't be written down mathematically? This will set the stage for the chapters to come.

1.1   Methods of Least Squares

Least-squares methods of parameter estimation (LSE) are so called because the goal is to choose the parameters that minimize the squared distance between the observations $y = \{y_1, y_2, \ldots, y_N\}$ and the predicted values given by the model. That is,

$$\text{SSE} = \sum_{i=1}^{N} (y_i - f(x_i, \theta))^2,$$

where N is the sample size and, as before, x represents the independent variables in the experiment, f is the model that relates x to y, and θ are the model parameters.[1]

As an example, simple linear regression assumes a model that predicts $y = mx + b$, so $f(x, \{m, b\}) = mx + b$. Minimizing SSE in simple linear regression yields the least-squares estimates $\hat{m} = r_{xy}\frac{s_y}{s_x}$ and $\hat{b} = \bar{y} - m\bar{x}$, where r_{xy} is the correlation between x and y, $\bar{x}$ and $\bar{y}$ are the sample means, and s_x and s_y are the sample standard deviations.

Simple linear regression is called simple for more than one reason: finding the least squares estimates $\hat{m}$ and $\hat{b}$ can be done by putting pencil to paper. In many situations the function *SSE* cannot be easily minimized, which requires that we use a computer program that searches for the minimum by proposing values for $\hat{m}$ and $\hat{b}$, computing the resulting *SSE*, and then, by proposing new values, attempting to make it smaller. There are many efficient algorithms to do this.

1.2   Maximum Likelihood

Maximum likelihood methods, the frequentist standard for parameter estimation, form the basis for many inferential procedures and model comparison methods. They rely heavily on optimization algorithms because the computations necessary for parameter estimation are usually more complex than those for least squares. In contrast to least squares, the distance between the data and the model's predictions is defined by how closely the probability distribution of the data matches the distributional assumptions of the model. This distance can be minimized by maximizing the likelihood of the data under the model.

Returning to the high-threshold recognition memory model, the data we observe are the numbers of "old" responses O_T to targets and O_D to distractors, out of a total number of items N_T and N_D presented in the experiment. We have already seen that

---

[1] We will use the notational convention that a variable name without subscripts such as y or x may be either vector or scalar valued; context should make clear which. If a variable is subscripted, such as y_i or x_i, it represents either an element of a vector or a scalar.

$$f(x, \{R, g\}) = \begin{cases} R + g(1 - R) & \text{if } x \text{ is old} \\ g & \text{if } x \text{ is new} \end{cases}$$

gives us the predicted proportions of "old" responses to target and distractor items. Using these predicted proportions, the probability distributions of O_T and O_D are binomial with parameters $\{N_k, p_k = f(k, \{R, g\})\}$ for $k = T, D$. The likelihood of the data $\{O_T, O_D\}$ is the product of the two binomial distributions, which is proportional to

$$\ell(\{g, R\} \mid \{O_D, O_T\}) = g^{O_D}(1 - g)^{N_D - O_D} [R + g(1 - R)]^{O_T}$$
$$\times [1 - (R + g(1 - R))]^{N_T - O_T} . \tag{1.1}$$

The maximum likelihood estimates of g and R are the values $\hat{g}$ and $\hat{R}$ that maximize $\ell(\{g, R\} \mid \{O_D, O_T\})$. This is another case where we can find $\hat{g}$ and $\hat{R}$ without complications and determine that

$$\hat{g} = O_D/N_D \text{ and } \hat{R} = \frac{O_T/N_T - O_D/N_D}{1 - O_D/N_D}$$

maximizes the function $\ell(\{g, R\} \mid \{O_D, O_T\})$. For more complicated likelihood functions, we will need numerical methods.

1.3 Bayesian Methods

Bayesian methods for parameter estimation do not just compute point estimates like least squares and maximum likelihood. As we described above, the final product of a Bayesian analysis is an estimate of a parameter's posterior distribution given the observed data. These methods incorporate the data's likelihood function, the same function used in maximum likelihood estimation, into Bayes' Theorem to arrive at this posterior distribution. Bayesian probabilities used to be called "inverse probabilities," a term that describes the problem of turning a likelihood into a probability distribution over parameters [22].

Bayes' Theorem is probably well known to most readers, but we will restate it here in terms of data, models, and parameters. A model, which we described earlier in terms of its predictions $f(x, \theta)$, states that data y will follow some probability distribution that is "tuned" according to its parameters θ. Using that probability distribution, we can write the likelihood L of y as a function of θ^2:

---

[2]Don't confuse the probability (or density) function $f_Y(y \mid \theta)$ with the model structure $f(x, \theta)$. The predictions of the model, described by $f(x, \theta)$ are not necessarily the same as the probability of the data given by $f_Y(y \mid \theta)$, though they were the same for the high-threshold model above. For the simple regression model, however, $f(x, \{m, b\}) = mx + b$, while most applications of

$$L(\theta \mid y) = \prod_{i=1}^{N} f_Y(y_i \mid \theta). \tag{1.2}$$

Although the likelihood L is a function of θ (where y is given), we can think of it (generally) as the probability (or the density) of the sampled measurements y given the parameters θ. Applying Bayes' Theorem, we want to invert the likelihood to obtain a probability (or density) of θ given the data y.

To do this, we will need to specify a prior distribution over θ. This distribution might reflect our past experiences with the model as it was fit to similar data (an informed prior), or we might choose to avoid making strong a priori assumptions about θ and instead choose an objective distribution that is uninformative or relatively flat. Such prior distributions usually spread probability over a wide range of possible parameter values. There are a number of different criteria by which an objective prior might be selected [23], but lack of information is probably the most popular basis for an objective prior.

The choice of a prior gives us a probability or density function $\pi(\theta)$ that represents the variability in the parameter θ before any data are observed. Bayes' Theorem states that

$$\pi(\theta \mid y) = \frac{L(\theta \mid y)\pi(\theta)}{f_Y(y)},$$

where $f_Y(y)$ is the marginal distribution of the data, taken over all possible parameter values. Because $f_Y(y)$ does not depend on θ, it is only a normalizing constant. It is usually very difficult to compute for models of any real complexity, and so we usually write

$$\pi(\theta \mid y) \propto L(\theta \mid y)\pi(\theta);$$

the posterior of θ given y is proportional to the product of the prior and the likelihood.

If we know $\pi(\theta \mid y)$ exactly, then we have everything we need to make inferences about the parameter θ. Not only can we compute point estimates (such as the posterior mean, mode, or median), we can compute exact probabilities for different hypotheses. We can evaluate the probability of a null hypothesis such as $H_0 : \theta \leq 0$, or construct the Bayesian equivalent of a 95% confidence interval: a credible set (θ_0, θ_1) such that $P(\theta \in (\theta_0, \theta_1)) = 0.95$.

Unfortunately, for most realistic models, we don't know $\pi(\theta \mid y)$ exactly, for one or two reasons. First, computing the normalizing constant $f_Y(y)$ is often complicated, preventing us from being able to write down closed-form solutions for $\pi(\theta \mid y)$. This problem, which was one major reason why Bayesian inference has lagged behind the development of frequentist techniques, has led to the development of algorithms that

---

regression state that y is normally distributed with mean $mx + b$ and some standard deviation σ. In this case, $f_Y(y \mid x, m, b, \sigma)$ is the normal density function that sketches out the bell curve.

permit us to sample values of θ from the posterior. These techniques, such as Gibbs sampling, Metropolis-Hastings sampling, Hamiltonian Monte Carlo sampling, and so forth, do not require explicit calculation of $f_Y(y)$, but instead approximate this marginalizing constant through Monte Carlo techniques.

The second reason we often don't know $\pi(\theta|y)$ exactly is because the likelihood function $L(\theta \mid y)$ may not have an explicit functional form. In psychology, neuroscience, and cognitive science, our goal is to develop a model that mimics human decision making, a process that is extremely complicated even for simple decisions. Often, while developing more complete explanations of behavioral data, models must grow in complexity to be able to account for different decision making patterns. For example, the high-threshold model may explain patterns of decisions from simple recognition memory experiments, but it is not equipped to handle more complex dynamics that appear in other memory experiments, such as those observed in free recall experiments [24, 25]. The benefit of more complex models is the power of unifying explanations for many different patterns of behavioral data at once, but the cost is usually one of computational complexity. It is often the case that as models become more complex, it becomes more and more difficult to determine the likelihood of the models' outputs with a set of equations.

And here lies the purpose of this book. There is a growing emergence of successful computational models in psychology, neuroscience, and cognitive science for which the likelihood functions are either unknown or computationally difficult to evaluate. Because the likelihood function has yet to be derived, one must explore the predictions of such models through simulations, and inference procedures are limited to the methods of least squares described above. In other words, due to complications in evaluating the likelihood function, the aforementioned computational models are unable to enjoy the many benefits that Bayesian analyses provide.

1.4 Approximate Bayesian Computation

There are now influential models in the behavioral sciences that are constructed from the "bottom up." Relatively well-understood neural mechanisms are quantified and used as the building blocks of more complex structures that can generate simple responses to quantitative representations of stimuli. Many of these models are used in memory and vision research. These models are tested by repeated simulation of the models' responses using constrained values of the parameters suggested by findings in neuroscience.

Fitting such models to data is orders of magnitude more demanding than the methods we have just outlined for models with explicit likelihoods. The most common method of estimating a simulation model's parameters is called approximate least squares [26, 27]. To understand approximate least squares, refer again to the high-threshold model of recognition memory. If we were to use approximate least squares, the parameter estimates would be obtained by first proposing reasonable values for R and g. These initial values would be used to simulate a number of responses to a sequence of target and distractor stimuli. For example, a "for" loop that cycles through the target items would first, by sampling from a Bernoulli distribution with probability parameter R_0, determine whether a trace had been laid down for each target item. All items with traces would be given an "old" response. All items without traces would then sample from another Bernoulli distribution with probability parameter g_0, and all items for which the sample was 1 would be given an "old" response. Another "for" loop would cycle through the distractor items, again sampling from a Bernoulli distribution with probability parameter g_0 to determine which distractors are given "old" responses. These two loops result in simulated values for $\hat{O}_{T,0}$ and $\hat{O}_{D,0}$, which can then be compared to the observed values O_T and O_D and evaluated as

$$\widehat{\text{SSE}}_0 = (\hat{O}_{T,0} - O_T)^2 + (\hat{O}_{D,0} - O_D)^2.$$

This would be the very first step in an optimization algorithm that would then select a new set of parameters $\{R_1, g_1\}$, perform a second simulation, the results of which would be used to compute $\widehat{\text{SSE}}_1$, and so on. Although the procedure is not difficult, it can demand enormous amounts of computing power to perform the simulation for each iteration, and, depending on the complexity of the problem, thousands of iterations may be necessary to find the optimal estimates of the model's parameters. Furthermore, because of the variability added by the simulated data, we shouldn't just simulate the data once for each proposed set of parameters, we really need to simulate the data many times, perhaps thousands of times, to reduce the influence of simulation variability on the value of SSE. However, the real reason that this approach is unsatisfactory is simply because it doesn't give us much information in the end: while we may have reasonably accurate point estimates for the parameters, we will not know how they are distributed, how they are correlated with each other, or what kinds of null hypothesis tests might be appropriate for determining if they are changing over experimental conditions.

Approximate Bayesian computation (ABC) was designed to overcome exactly this kind of problem. Originally developed by Pritchard et al. [28], ABC proceeds in a way similar to approximate least squares, replacing the computation of the likelihood with a simulation step. The simulation step produces a sample of simulated data X that is evaluated relative to the observed data Y. This evaluation is made on the basis of the distance between X and Y, and distance can be defined in a number of ways. The SSE is one example of a distance, in which the samples X and Y could be represented by sample statistics like their means and variances. However, ABC does not use distance minimization to generate point estimates of parameters, but rather to estimate the posterior distributions of the parameters.

The logic behind ABC is the following: if a proposed parameter value θ^* is able to generate a simulated data set X that is close to the observed data Y, then it must have associated with it a relatively high posterior probability. Therefore, for some distance function $\rho(X, Y)$, we will keep all values of θ^* that result in $\rho(X, Y) \leq \epsilon_0$ and discard the rest. If we choose $\rho(X, Y)$ and ϵ_0 correctly, then $\pi(\theta \mid \rho(X, Y) \leq \epsilon_0)$ will approximate $\pi(\theta \mid Y)$ [28].

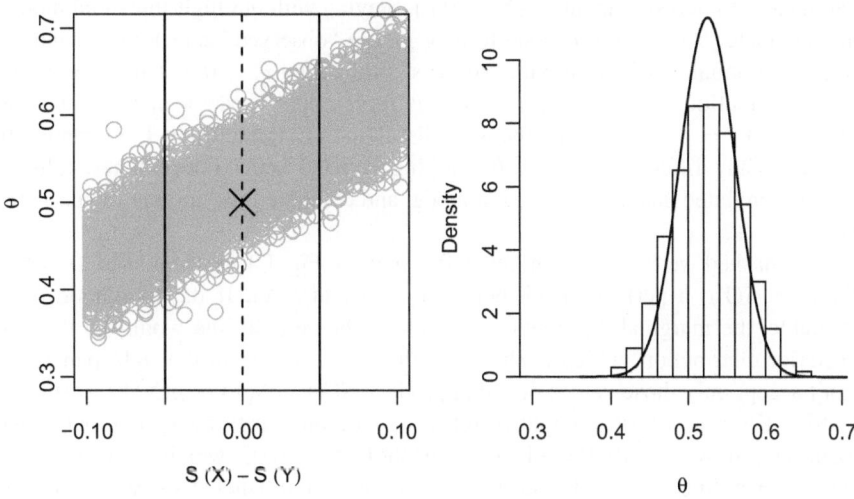

Fig. 1.1 Intuition behind approximate Bayesian computation. The left panel shows the joint distribution of the parameters of interest θ against the distance between the statistics of the observed data $S(Y)$ and the simulated data $S(X)$. The dashed vertical line represents the case where $S(Y) = S(X)$, and the solid black lines represent the degree of tolerance ϵ. The right panel shows the estimated posterior distribution (histogram) under the level of ϵ in the left panel, overlaid by the true posterior (black density)

Consider one last time the high-threshold model of recognition memory. Let

$$\widehat{\text{SSE}} = \left(\hat{O}_T^* - O_T\right)^2 + \left(\hat{O}_D^* - O_D\right)^2$$

be the distance function, where $\hat{O}_T^*$ and $\hat{O}_D^*$ are the simulated data generated by proposed parameter values $\{R^*, g^*\}$. We need to make sure that the number of simulated trials $N_D + N_T$ is the same as the number of trials in the experiment to ensure that the sampling distributions of $\hat{O}_T^*$ and $\hat{O}_D^*$ are comparable to those of O_T and O_D. If $\widehat{\text{SSE}}$ is less than ϵ_0, then we keep $\{R^*, g^*\}$ as a sample from the posterior. If it is greater than ϵ_0 we discard it and sample a new $\{R^*, g^*\}$, possibly from the prior or from some other proposal distribution, and repeat the simulation and computation of $\widehat{\text{SSE}}$. How the proposals are sampled and how ϵ_0 changes (or not) with repeated sampling are determined by the specific ABC algorithm that we choose for this particular problem. We will discuss these algorithms later in Chap. 2.

Figure 1.1 illustrates the logic of the ABC approach more generally. Let $S(X)$ and $S(Y)$ be functions that produce summary statistics (means, variances, quantiles, etc.) of the simulated data X and the observed data Y. For example, the statistics could be the number of "old" responses to target and distractor items for the observed (i.e., O_T and O_D, respectively) and simulated (i.e., $\hat{O}_T^*$ and $\hat{O}_D^*$) data. The distance function $\rho(X, Y)$ is $|S(X) - S(Y)|$. The left panel plots the joint distribution of the parameter of interest θ against $S(X) - S(Y)$. Staying with our high-threshold model, the parameter θ could correspond to R or g. The "observed" data Y were sampled from a binomial distribution with "success" probability $\theta = 0.5$. This point in the joint distribution of θ and $S(X) - S(Y)$ is represented by an $\times$ at $\theta = 0.5$ and $S(X) - S(Y) = 0$. To generate the joint distribution in the left panel, we randomly selected many different values for θ^* ranging from 0.3 to 0.7. For each new value of θ^*, we simulated data from binomial model and computed the number of successes $S(X)$.

The dashed vertical line in the left panel of Fig. 1.1 is located at 0, when $S(X) = S(Y)$: a perfect match between $S(Y)$ and $S(X)$. If the likelihood were available, the marginal distribution of θ along the vertical line would be the true posterior distribution, which is shown as the black density in the right panel. We can't accept only those values of θ that produce $S(X) = S(Y)$ or $\rho(X, Y) = 0$; such a strict distance criterion would result in an extraordinarily heavy computational load. Instead, we specify the tolerance threshold $\epsilon_0 = 0.05$, which is shown as the solid vertical lines to the left and right of zero in the left panel. This value of ϵ_0 lets us retain enough samples of θ to be able to construct a relatively accurate estimate of θ's posterior. The right panel of Fig. 1.1 shows the histogram of the values of θ in the left panel that produced simulated data X such that $|S(X) - S(Y)| < \epsilon_0$; this is the region of the joint distribution that falls between the two solid vertical lines. The histogram estimate is close to the true posterior that would be obtained had a likelihood been known.

More generally, the relationship between the marginal posterior distribution of a parameter θ and the joint distribution of that parameter and the distance $S(X) - S(Y)$ shown in the left panel of Fig. 1.1 can be expressed as

$$\pi(\theta \mid Y) \propto \int_{\mathcal{X}} \pi(\theta) f(x, \theta) I(\rho(X, Y) \le \epsilon) \, dx, \tag{1.3}$$

where $\mathcal{X}$ is the support of the simulated data and $I(a)$ is an indicator function returning one if the condition a is satisfied and zero otherwise. The integration in Eq. (1.3) expresses the marginalization over the random variable $\rho(X, Y)$ that was performed to provide an estimate of θ in the right panel of Fig. 1.1. All values of θ producing data that fell within the black vertical lines were accepted. Note that this marginalization does not take into account the obvious trend in the relation of θ to $S(X) - S(Y)$; this is an important aspect of some versions of ABC algorithms that we will discuss in the next chapter.

1.5   Outline

The focus of this book is to illustrate a variety of ABC techniques on psychological problems. As such, while we will review many different types of ABC algorithms, we will highlight a set of algorithms that have been developed for particular situations that arise regularly when doing cognitive modeling. In the next chapter, we will outline several different ABC algorithms, focusing in particular on those approaches most similar to the ones we advocate for psychological models. This is not intended to be an exhaustive review of ABC algorithms. Interested readers may consult [29–35] for reviews, additional options, and more mathematical background. In the third chapter, we provide a worked example on the Minerva 2 model [36]. For this model, we provide simulations using two of the algorithms described in the second chapter, and compare their accuracy to a set of analytic expressions describing the limiting behavior of the model [37]. In the fourth and fifth chapters, we discuss a number of applications of ABC algorithms on interesting problems in psychology. In the sixth and final chapter, we provide an outlook on the ability of ABC techniques to advance the field of cognitive science, and discuss the role of mathematical tractability in the development of psychological theory.

Likelihood-Free Algorithms

2

2.1 Introduction

In this chapter we present the technical details behind several algorithms for performing likelihood-free inference. The algorithms vary along a number of dimensions, which makes organizing them difficult. At the most basic level, the likelihood-free algorithms we discuss in this book can be described roughly in five essential steps:

1. Generate a candidate parameter value θ^*.
2. Generate a set of simulated data X using θ^* and a model.
3. Summarize the properties of X.
4. Compare the summary of X to a summary of Y.
5. Apply a weight to θ^* that reflects how close X is to Y.

Across the literature, most likelihood-free algorithms have focused on developing innovative techniques on either Steps 1, 3, or 5. Regarding Step 2, most likelihood-free algorithms agree that simply generating a large set of simulated data X should be good enough as long as the summaries in Step 3 are relatively stable. As such, while we don't consider Step 2 to be an important step as far as classifying the algorithms is concerned, because it clearly interacts with the other steps, close attention to this step is warranted. For example, determining the size of X will depend on how quickly one can simulate data from the model, as well as how large the observed data Y are. Furthermore, some algorithms that we describe below require much larger sets of simulated data to ensure a quality likelihood approximation can eventually be formed.

Regarding Step 4, most likelihood-free algorithms specify that a simple Euclidean distance can be used to compare the statistics of the simulated data $S(X)$ to the statistics of the observed data $S(Y)$. In other words, we can simply

© Springer International Publishing AG 2018
J.J. Palestro et al., *Likelihood-Free Methods for Cognitive Science*,
Computational Approaches to Cognition and Perception,
https://doi.org/10.1007/978-3-319-72425-6_2

calculate $S(X) - S(Y)$ to evaluate how near X is to Y. However, again we find that the steps listed above interact with one another: As Step 4 depends on the way in which the data are summarized via $S(\cdot)$, Step 4 highly depends on Step 3. Some algorithms depart in the way the data X are summarized [38], and so they also differ in how X is related to Y. As we describe below, one way of describing the distribution of X might be through kernel density estimation [39] or even a histogram method [40]. In this case, as the distribution of X has been specified as a proper probability density function, we can evaluate the probability of observing the data Y under the distribution of X.

Although the steps listed above may give the impression that all likelihood-free algorithms are simple, this is unfortunately not the case. Many sophisticated techniques have been created in the hopes of increasing the efficiency of an algorithm on a given problem, and as one might expect, the efficiency of the algorithms below do vary by the type of problem to which they are applied. Because the algorithms we present later in this chapter are sometimes complex, we first introduce a few concepts at a high level by describing the different choices one can make at Steps 1, 2, or 3.

2.1.1 Generating Candidate Parameter Values

As described in Chap. 1, our ultimate goal is to assemble a collection of samples from the desired posterior distribution $\pi(\theta \mid y)$. To achieve this goal, the algorithms of this chapter generate a sequence of candidates θ^* and then either add them to the collection of posterior samples or discard them entirely. The more times candidates θ^* can be generated and accepted into our collection, the more efficient our sampler will be. As such, an important consideration in likelihood-free algorithms is how to generate effective proposals.

For example, when generating candidate parameter values at Step 1, we might elect to simply draw a random sample from the prior so that $\theta^* \sim \pi(\theta)$. This strategy can be effective computationally because the prior can often be chosen such that it is easy to draw samples from. Furthermore, when the prior is chosen such that it is of a similar shape and location to the desired posterior, more candidate proposals θ^* will resemble the posterior, which also increases the computational efficiency. However, when the details of the posterior are not well known, it can be inefficient to sample from a distribution (i.e., the prior) that is very different from our posterior. In this case, we might choose to develop a recursive strategy where the quality of previous candidates in our sampling algorithm can be used to guide the candidate generation process.

Basing proposal samples off of previous states is precisely the strategy behind conventional Markov chain Monte Carlo methods. In these methods, the procedure is to first initialize a "chain" by picking a value θ_0 that has some nonzero density in the posterior distribution. On the first iteration, we sample a new candidate value from a distribution that depends on the initial value θ_0. Depending on the properties of this sampling distribution, the new proposals will be more or less similar to the initial value. To evaluate whether the new proposal is better or worse than the initial value, we rely on the Metropolis-Hastings rule, which we discuss below. If the new proposal is deemed better, the chain moves to this new place in the posterior; otherwise, the chain stays put at the initial value. The procedure continues until enough samples have been drawn to form a good approximation of the posterior.

The Markov chain strategy is typically more effective than simply sampling from the prior distribution, as the chain has the ability to move nearer to the high-density regions of the posterior when the prior is different from the posterior. However, it does have a sampling dependency issue where new samples are highly correlated with previous states of the chain. This feature of the posterior samples is known as "autocorrelation," and can have undesirable effects on the approximated posterior. Furthermore, if the posterior has multiple modes, a chain can get stuck in a local mode of the posterior and be unable to explore the rest of the parameter space.

One easy solution to the aforementioned problems is to run multiple chains simultaneously. Here, the likelihood of all the chains getting stuck in a local mode is inversely proportional to the number of chains. Better still is to use the information contained in the distribution of current states of the chains to generate new proposals. This strategy is at the heart of so-called particle filtering approaches, where chains are referred to as "particles," and the algorithms work to evolve a set of particles from the prior to posterior. Often, the particles share distributional information with one another so that proposal generations can automatically "tune" themselves to be more efficient. For example, if the particles have a wide distribution, it might suggest that the posterior is also wide, and so a wider distribution should be used to generate the candidate proposals.

While we will talk more about the mathematical details of these three proposal strategies below, hopefully the discussion above hints to the idea that different strategies are essential for different types of problems. For example, simple problems could be approached by sampling proposals from the prior, whereas more difficult problems require sophisticated techniques like particle filtering. In the end, the user will need to take these features of the problem into account when choosing among the likelihood-free algorithms we have discussed here.

2.1.2 Summarizing the Simulated Data

Another consideration comes at Step 3, where one must decide how to summarize the distribution of simulated and observed data. Typically, it is too computationally costly to use the entire set of data, and so the conventional approach in these likelihood-free techniques is to calculate some statistics of the data. Ideally, the statistics computed from the data characterize the distribution of data properly, so that parameters can be estimated accurately. In other words, our task is to choose a set of statistics that contain as much information about the unknown model parameters as the entire set of data itself. This property is known as *sufficiency*.

The property of sufficiency stipulates that a sufficient statistic $S(\cdot)$ of some data Y provides as much information about the data when estimating a model's parameters

θ as does the whole data set itself. Thus, if $S(Y)$ is a sufficient statistic for the parameters θ, then the posterior distribution can be written as

$$\pi(\theta \mid Y) = \pi(\theta \mid S(Y)).$$

Sufficiency, therefore, is something that depends on the structure of the model—the model's likelihood. This key assumption of sufficiency is often overlooked, but it is critical for obtaining accurate posterior estimates. To determine if a statistic $S(Y)$ is sufficient, we must be able to reexpress the likelihood $f(y|\theta)$ as a function of the sufficient statistic and the data. By the Fisher-Neyman Factorization Theorem [41], if $f(y|\theta)$ can be factored as

$$f(y|\theta) = g\left(S(y)|\theta\right) h(y) \tag{2.1}$$

for some function $h(y)$ then $S(y)$ is sufficient for the parameter θ.

As an example, consider a series of n independent and identically-distributed Bernoulli trials and let $Y_i \in \{0, 1\}$ be the outcome on trial i. We can write

$$P(Y_i = y) = \begin{cases} \theta^y(1-\theta)^{1-y} & \text{for } y \in \{0, 1\} \\ 0 & \text{otherwise,} \end{cases}$$

where $\theta \in [0, 1]$ is the probability that $Y_i = 1$. The joint probability function for the set of outcomes $\{Y_1 = y_1, Y_2 = y_2, \ldots, Y_n = y_n\}$ is

$$f(y|\theta) = \prod_{i=1}^{n} P(Y_i = y_i)$$

$$= \prod_{i=1}^{n} \theta^{y_i}(1-\theta)^{1-y_i}$$

$$= \theta^{\sum_{i=1}^{n} y_i}(1-\theta)^{n-\sum_{i=1}^{n} y_i}$$

$$= \left(\frac{\theta}{1-\theta}\right)^{\sum_{i=1}^{n} y_i} (1-\theta)^n.$$

Therefore, the function $f(y|\theta)$ can be written as a function of the unknown parameter θ and the statistic $S(y) = \sum_{i=1}^{n} y_i$. By Eq. (2.1), we can let $g(S(y)|\theta) = g\left(\sum_{i=1}^{n} y_i|\theta\right)$ and $h(y) = 1$, demonstrating that the statistic $S(y) = \sum_{i=1}^{n} y_i$ is sufficient for the parameter θ.

For our purposes, requiring that $S(Y)$ be sufficient is problematic because we can't examine an unknown or intractable likelihood to determine if the Fisher-Neyman Factorization Theorem holds. Although there are a growing number of different strategies for resolving the sufficiency problem [42], the most common approach has been to select a large set of summary statistics and hope that this set of statistics is large enough that it is at least close to being jointly sufficient for the parameters of interest. While adding more summary statistics to this set will tend to provide more information about θ, sufficiency can still never be guaranteed, and some summary statistics may provide identical information about the model parameters. When a set of summary statistics are not sufficient for the parameters, then the influence of the information conveyed by the observed data will be weaker, resulting in posterior distributions that are inaccurate, particularly with respect to the degree of variability in the estimated posteriors [29].

Clearly, deciding how to characterize the data is an important choice in the likelihood-free context. Namely, it is impossible to know whether or not a set of statistics is sufficient for the unknown parameters if the likelihood is intractable. Given this alarming fact, other approaches, such as the probability density approximation method [38], approximate the distribution of data itself. The idea behind these algorithms is that the data are always sufficient to themselves, and so the issue of sufficiency is solved. Unfortunately, as we will discuss below, while these algorithms free us from one problem, they introduce another problem in the form of computational complexity. Because the data are not summarized by statistics, these algorithms rely on extensive model simulations to form an accurate approximation of the posterior. Again, the choice of whether to calculate sufficient statistics, groups of summary statistics, or use the entire data set depends on many factors, specifically computational speed. The user must balance these factors when choosing among the set of algorithms we discuss below.

2.1.3 Weighting Strategies

The final major consideration in choosing among likelihood-free algorithms is how to evaluate the quality of the candidate proposals θ^*. The best way to think about this problem is through the joint distribution of the proposals θ^* and the distribution of the simulated data X associated with them. First, let's assume that the statistic $S(\cdot)$ properly characterizes the observed data Y and the simulated data X. Because the observed data Y have already been observed, they are assumed to be known and will not change as new parameter proposals are considered. Hence, the statistics $S(Y)$ will also stay constant. The only random variables to consider now are θ^* and the associated $S(X)$. The left panel of Fig. 1.1 shows one example of a joint distribution of $(\theta^*, S(X))$. Here, the figure shows that the distance between the observed data and the simulated data is a linear function of the parameter proposals: larger distances of $S(X) - S(Y)$ are associated with parameter proposals that are farther from the true data generating value of $\theta = 0.5$.

The left panel of Fig. 1.1 is useful because it shows how these two variables are related and how the concept of proposal "fitness" can be incorporated into our evaluation. Perhaps the simplest way to evaluate a candidate proposal would be a binary pass/fail decision where proposals that produced data that were close enough to the observed data would be kept and proposals that did not would be rejected. But, how would we define "close enough?" One can imagine adopting a simple tolerance threshold ϵ that could be used to apply a cutoff for the distances $|S(X) - S(Y)|$. For example, the right side of Fig. 1.1 shows what the distribution of θ^* values are such that $|S(X) - S(Y)| < \epsilon = 0.5$. This strategy forms the core of rejection-based ABC algorithms.

While adopting a simple pass/fail rule is simple, it is exceedingly wasteful depending on the value of ϵ. For example, in producing the posterior in the right panel of Fig. 1.1, half of the simulated data were simply discarded. As an alternative, consider applying a different function that was not piecewise pass/fail. Instead, consider applying a *kernel* to the distribution of $S(X) - S(Y)$ in order to obtain "weights" for each associated θ^*. We could choose something symmetric such that large values of $S(X) - S(Y)$ would result in smaller weights for θ^*, and small values (i.e., near zero) would result in larger weights. Suppose further that the sharpness of the gradient associated with the proposal weights could be increased or decreased with a parameter δ, much in the same way as ϵ in the rejection-based ABC methods. Specifically, increasing δ would allow more proposals to have relatively large weights, whereas decreasing δ would cause some proposals to have large weights (i.e., those generating data X such that $S(X) - S(Y)$ was near zero) and most other proposals to have small weights. Clearly, the accuracy of the estimated posterior would be intimately interwoven with the choice of δ; but, assuming we could make good choices about δ, we could arrive at accurate posteriors. This type of weighing strategy is what makes kernel-based ABC algorithms so effective: they can arrive at similarly accurate posterior estimates without throwing away large percentages of the data.

While kernel-based ABC approaches work well in using the entire set of simulated data, they require a functional form for the kernel, and this choice is not always straightforward. While most kernels are selected to be symmetric, unimodal, and have exponentially decreasing weights as $|S(X) - S(Y)|$ increases, beyond these features, kernels are chosen in a somewhat subjective and arbitrary manner. Sometimes, the choice of kernel can have adverse effects on the quality of the posterior estimates, especially in cases where the distribution of summary statistics is substantially different from the kernel function. A related issue centers on the choice of δ: when δ is too large, the posterior estimate is too wide relative to the true posterior, and when it is too small, sampling methods such as MCMC and particle filtering are more difficult to use effectively.

As a solution to the issues associated with kernel-based ABC methods, other approaches such as the synthetic likelihood method [43], have exploited facts about the distribution of summary statistics to avoid the specification of the kernel altogether. Specifically, the synthetic likelihood method relies on the central limit theorem to specify the distribution of statistics on summary statistics. The idea is that, regardless of the set of summary statistics chosen, if enough new data sets are generated, one can construct a distribution of summary statistics, and the mean of these summary statistics is normal in shape according to the central limit theorem. Using this two-level approach, the synthetic likelihood method escapes the complications of specifying a kernel as well as choosing the parameter δ. However, similar to the probability density approximation method, the synthetic likelihood method requires extra simulations that make it computationally costly to implement.

2.1.4 Outline of the Chapter

Although the steps used to implement likelihood-free samplers seem simple, hopefully the discussion above illustrates that there are many considerations when choosing an algorithm for a specific problem. The set of choices one can make in the steps above correspond to certain constellations of algorithms, and some of these constellations have been developed, tested, and named by likelihood-free researchers. As a road map of which algorithms are suitable for different problems, Table 2.1 shows how the algorithms compare on a number of different features. For example, all of the algorithms are flexible with the method of proposal generation (i.e., the first row), yet all of the algorithms use a unique weighting approach (i.e., the fourth row). The rows of Table 2.1 roughly correspond to the generic algorithmic steps from the introduction: the first row corresponds to Step 1, the second row corresponds to Step 3, the third row corresponds to Step 4, and the fourth row corresponds to Step 5. The final row considers another important feature: dimensionality. As rejection-based algorithms are wasteful and synthetic likelihood and probability density approximation methods are computationally costly, they are not as practical as kernel-based methods when the number of dimensions is high. However, as we will discuss below, these four classes of algorithms can actually be combined with hierarchical approaches to move beyond low-dimensional problems when dealing with hierarchical models. In sum, when choosing among likelihood-free algorithms to fit a computational model to data, we recommend using Table 2.1 as a guide for specific applications.

Organizing the various likelihood-free algorithms is not an easy task, and there are probably several effective ways to organize and classify them. For our purposes, we found it convenient to classify the algorithms on the basis of Step 5 above, which is mostly closely related to how the likelihood function is approximated. Given our choice of organization, we formed four categories of algorithms and present them in the following order: rejection-based, kernel-based, general methods, and hierarchical methods. Within each class of algorithm, we will discuss a number of

Table 2.1 Some considerations when choosing a likelihood-free algorithm

Consideration	RABC	KABC	SL	PDA
Is it flexible with proposal generation?	Yes	Yes	Yes	Yes
Does it require sufficient statistics?	Yes	Yes	Yes	No
Does it use error terms?	Yes	Yes	No	No
What type of weighting approach is used?	Binary	Gradient	CLT	KDE
Is it suitable for high dimensions?	No	Yes	No	No

The first column proposes a question for consideration along several dimensions. Columns 2–5 explain how each of the four core likelihood-free algorithms matches up to the dimension under consideration

RABC rejection-based approximate Bayesian computation, *KABC* kernel-based ABC, *SL* synthetic likelihood, *PDA* probability density approximation, *CLT* central limit theorem, *KDE* kernel density estimate

interesting alternatives, some of which only vary in the way they generate proposals during the sampling process. It is important to keep in mind that different choices can be made at different steps in the development of an algorithm, and so while some algorithms have different names, they are often strikingly similar. We now discuss each of the four classes of algorithms in turn.

2.2   Rejection-Based ABC

Rejection-based samplers, discussed briefly in the previous chapter (see Fig. 1.1), are the simplest of all the algorithms, and their simplicity is both a blessing and a curse. These algorithms are typically very easy to program, but they also are the most inefficient. The algorithms work in the following way: For the parameter(s) of interest θ, a proposal θ^* is drawn at random from a proposal distribution (e.g., the prior distribution $\pi(\theta)$). Each proposed θ^* is then used to simulate data X from the model, so $X \sim \text{Model}(\theta^*)$.[1] To decide whether θ^* comes from the desired posterior $\pi(\theta\,|\,Y)$, we compare the simulated data X to the observed data Y by way of a discriminant function $\rho(X, Y)$. In rejection-based approaches, if the simulated data X is similar enough to Y, producing a $\rho(X, Y)$ less than some tolerance threshold ϵ, then we assume that the proposed θ^* must have some nonzero probability of coming from the posterior distribution $\pi(\theta\,|\,Y)$. Hence, we accept θ^* as a sample from the posterior. Otherwise, if $\rho(X, Y)$ is too large (i.e., $\rho(X, Y) > \epsilon$), we discard θ^* and repeat the process until a desired number N of samples have been obtained. We call this type of sampling *rejection-based ABC*, and the interested reader should consult [44] for more details.

Figure 2.1 provides pseudocode for a typical rejection-based sampler. It is interesting that such a simple algorithm (fewer than 10 lines of code; much simpler

---

1: Given data and model $Y \sim \text{Model}(\theta)$, tolerance threshold ε, and prior distribution $\pi(\theta)$:
2: **for** $1 \leq i \leq N$ **do**
3:     **while** $\rho(X,Y) > \varepsilon$ **do**
4:         Sample θ^* from the prior: $\theta^* \sim \pi(\theta)$
5:         Generate data X from θ^*: $X \sim \text{Model}(\theta^*)$
6:         Calculate discrepancy $\rho(X,Y)$
7:     **end while**
8:     Store $\theta_i \leftarrow \theta^*$
9: **end for**

Fig. 2.1 An ABC rejection sampling algorithm to estimate the posterior distribution of a parameter θ given data Y

---

[1]The notation $\text{Model}(\theta)$ describes the distribution of a random variable X, whereas the notation $\text{Model}(y|\theta)$ denotes the probability density at the location y, conditional on the parameters θ, as in Eq. (1.2).

than most traditional likelihood-based samplers) can provide a reasonably accurate estimate of the joint posterior distribution (see Fig. 1.1). While Fig. 2.1 provides generic pseudocode that can be adopted for any programming language, we can also provide explicit code to run a simple rejection-based ABC sampler within R. First, suppose we have a set of data Y that come from some experiment. For the sake of illustration, suppose these data are accuracy measures on some task, where a correct answer on trial i would be coded $Y_i = 1$, and an incorrect response would be coded $Y_i = 0$. First, to generate these data, we can run the following block of code:

```
p=.5              # true value for p
n=100             # number of trials
data=rbinom(n,1,p)       # generate data
```

Here, Line 1 declares the true value for the unknown parameter p that we will estimate for this problem. The variable n sets the number of observations in our data, which correspond to the number of trials from our experiment. Finally, Line 3 simulates some random data from a Binomial model using the parameter p. Hence, the variable data are the data Y from our experiment.

In the real world, we would not know what the true value of p is, and so we would want to estimate this parameter in order to perform inference on our data. Furthermore, in real-world scenarios, we won't know what the true data-generating model is; instead, we must make an assumption about the model that might have conceived of the data we observed. Having obtained the data Y, suppose we assumed that the data could be described by a Binomial distribution such that

$$Y \sim \text{Binomial}(n, p),$$

where n is the number of trials, and p is the probability of observing a "success" (i.e., $Y_i = 1$) on a single trial. It happens that the model we are assuming generated the data actually did generate the data for the sake of illustration, but of course this will not generally be the case. Because we have assumed a Binomial model, the probability density function can be easily defined, and the likelihood function could be written

$$\mathscr{L}(p \mid Y) = \binom{n}{k} p^k (1-p)^{n-k}, \tag{2.2}$$

where the term $k = \sum_{i=1}^{n} Y_i$ represents the total number of successes observed across the n trials. To obtain the posterior distribution, we could specify a prior for the parameter p, such as a uniform distribution across the entire unit interval $(0, 1)$. One way to do this would be to specify a Beta distribution as the prior for p, where under some values of the parameters, the Beta distribution becomes the uniform distribution. Specifically, we could set

$$\pi(p) = \text{Beta}(\alpha = 1, \beta = 1),$$

where α is the shape parameter and β is the scale parameter. Under this setting, the Beta distribution has equal probability across the unit interval. Hence, the posterior we wish to obtain is

$$\pi(p|Y) \propto \mathscr{L}(p|Y)\pi(p). \tag{2.3}$$

Because the posterior in Eq. (2.3) is tractable, we could use any number of conventional algorithms to draw samples and approximate the posterior. In fact, this particular posterior is fully analytic, meaning that we wouldn't need to approximate it at all. However, as this example is purely for illustrative purposes, suppose that the likelihood function in Eq. (2.2) was intractable, and we were forced to approximate it using one of the many algorithms appearing in this book. Suppose, after consulting Table 2.1, we chose a simple rejection-based ABC approach for this problem. One important step would be to decide on a summary statistic and a tolerance threshold. From our discussion about sufficiency above, as well as from Fig. 1.1, we might choose to evaluate the probability of success across all trials, so that

$$S(Y) = \frac{1}{n} \sum_{i=1}^{n} Y_i.$$

Then, to compare the summary statistics of the observed data Y to the simulated data X (see Fig. 2.1), we might choose

$$\rho(X, Y) = |S(X) - S(Y)| = \frac{1}{n} \left| \sum_{i=1}^{n} X_i - \sum_{i=1}^{n} Y_i \right|. \tag{2.4}$$

Finally, we should define a tolerance threshold such that when $\rho(X, Y)$ is calculated, it can be tested for quality (i.e., Step 5). To set these functions up in R, we can run the following block of code:

```
eps=.05              # tolerance threshold
N=1000               # number of particles
theta=numeric(N)     # declare a vector for storage

p.alpha=1   # prior for first parameter of the beta distribution
p.beta=1    # prior for second parameter of the beta
     distribution

rho=function(x,y)  abs(sum(x)-sum(y))/n      # rho function
```

Line 1 sets the value of eps corresponding to ϵ, Line 2 sets the total number N of samples to be drawn from the posterior, and Line 3 sets up a storage object theta to collect all of the accepted samples. Lines 5 and 6 specify the shape of the prior distribution, and Line 8 specifies a function in R that will calculate $\rho(X, Y)$ from Eq. (2.4).

With all of our variables initialized and the details of our algorithm fully specified, we need only convert the pseudocode in Fig. 2.1 to R code to draw samples from the desired posterior. The next block of code will perform this step continuously until N samples have been drawn:

```
for(i in 1:N){              # loop over particles
    d=eps+1                 # initialize d to be greater than eps
    # continue proposal generation until condition is satisfied
    while(d>eps) {
    theta.1=rbeta(1,p.alpha,p.beta)  # sample from prior
    x=rbinom(n,1,theta.1)            # simulate data
    d=rho(data,x)                    # compute distance
    }
    theta[i]=theta.1                 # store the accepted value
}
```

Lines 1 and 10 begin and end the for loop over the number of samples. Embedded within the for loop is a while loop (Lines 4 and 8) that will continue testing new proposals until a candidate is accepted. To force a continuous while loop, Line 2 specifies that the evaluation variable d is initialized to be a value greater than the tolerance threshold variable eps. Within the while loop is the process we have already described for rejection-based samplers: draw a sample from the prior (Line 5), generate synthetic data (Line 6), and evaluate its degree of match to the observed data (Line 7). For each new proposal, the variable d is reset, but its value is irrelevant as long as d is greater than eps. Once d is small enough, the proposal value theta.1 is stored into the vector theta, and the process restarts until N samples have been taken.

Once the blocks of code above have been pasted into your R console, obtaining samples from the posterior should be relatively quick (i.e., should take around 10 s under these initial settings). While this example is very simple, one can gain a sense of the power of this approach in other, more realistic situations, considering the fact that the inference procedure was driven purely by model simulation. In other scenarios, such as when the data are difficult or time consuming to generate from the model, we recommend coding the model up in compiler languages like C or Fortran to improve the sampling speed. Once this is done, one only needs to change Line 6 to call the compiled code from R. Of course, the efficiency of the sampler we have presented here depends on several other factors, such as the amount of data we have (i.e., the variable n), the number of samples we wish to collect from the posterior (i.e., the variable N), how close the prior is to the posterior (i.e., which can be manipulated by choosing different values for the variables p.alpha and p.beta), and the tolerance threshold ϵ (i.e., the variable eps). Building on our discussion from Chap. 1, choosing an appropriate value for ϵ can be a difficult problem, because we don't know for any particular model what the optimal value of ϵ is. As a general rule, we would like ϵ to be small because, as ϵ approaches zero, the approximation becomes exact [29]. However, requiring smaller values of ϵ will produce longer computation times [44]. Because the choice of ϵ is such an important one, we will revisit this issue later. We now turn to specific rejection-based algorithms: the rejection-based MCMC algorithm and a number of sequential Monte Carlo algorithms.

2.2.1   The Rejection-Based MCMC Algorithm

As discussed in the introduction of this chapter, it can be inefficient to draw proposal values θ^* from the prior distribution when the prior is markedly different from the posterior. While in the illustrative example above the posterior was similar to the prior, in practice, this is often not the case. Furthermore, we rarely know what the shape and location of the posterior will be relative to the prior, and so evaluating the efficacy of the algorithm in Fig. 2.1 become a trial-and-error process.

Fortunately, several algorithms have been developed to improve the efficiency and robustness of proposal generation. One example is the class of ABC MCMC algorithms, which use conventional MCMC algorithms to generate proposals for θ. These MCMC algorithms, which have been instrumental in advancing Bayesian estimation techniques [45, 46], are iterative procedures that evaluate and filter parameter proposals θ^* in such a way that we eventually obtain a sample of proposal values that are from the desired posterior distribution. The most popular MCMC sampler is the Metropolis-Hastings algorithm [46].

To perform Metropolis-Hastings sampling, we begin by first initializing a "chain" of parameter values with some initial value θ_0. Given θ_0, we then sample a candidate value θ^* obtained from a proposal distribution, such as a Gaussian distribution, from which sampling is easily done. We denote the proposal distribution as $q(\theta^*|\theta_0)$. The distribution of the first proposal θ^* is conditioned on the initial value θ_0, but for subsequent proposals, their distributions will be conditioned on the current value of the chain. In our example, if we assume the proposal distribution is Gaussian, we would write

$$\theta^* \sim \mathcal{N}(\theta_0, \sigma),$$

where $\mathcal{N}(a, b)$ denotes a Gaussian distribution centered at the value a with standard deviation b. The standard deviation σ of a proposal distribution is referred to as a "tuning parameter," because it controls the range over which proposals θ^* can be sampled.

To evaluate the quality of the newly proposed value θ^*, we must calculate its relative posterior density and compare it to the relative posterior density of the initial value θ_0. If the posterior density of θ^* is higher than that of θ_0, we keep it and it becomes the next value θ_1 in the chain. If it is lower, we accept the value of θ^* with a probability determined by the ratio of the posterior densities. In this way, the values in the chain tend to move toward the highest probability regions of the posterior, but can still move away from these high-probability regions and toward the tails of the posterior.

To perform Metropolis-Hastings sampling in the likelihood-free context, we calculate the probability of accepting θ^* by evaluating

$$\alpha = \begin{cases} \min\left(1, \dfrac{\pi(\theta^*)q(\theta_0|\theta^*)}{\pi(\theta_0)q(\theta^*|\theta_0)}\right) & \text{if } \rho(X, Y) \leq \epsilon \\ 0 & \text{if } \rho(X, Y) > \epsilon, \end{cases} \tag{2.5}$$

where $\pi(\theta)$ is the prior distribution for θ and q is the proposal distribution. To implement the sampler in a computer program, after computing α for θ^*, we would draw a sample from a uniform $[0, 1]$ distribution. If the sampled value is less than α, we accept θ^* as the new value of the chain by setting $\theta_1 = \theta^*$. However, if the sample value is greater than α, we reject the proposal and set $\theta_1 = \theta_0$, and so the chain does not move.

What makes Eq. (2.5) different from a standard Metropolis-Hastings acceptance probability is that the value of α is determined by whether $\rho(X, Y) > \epsilon$. For the proposal θ^* to have any chance to be accepted as a sample from the posterior distribution it must first generate data X that is within ϵ of the true data Y. The procedure iterates, like the standard Metropolis-Hastings algorithm, replacing θ_0 with θ_{t-1} in Eq. (2.5) to evaluate the quality of the proposal θ^* on iteration t, until we have obtained a chain of m values $\{\theta_0, \theta_1, \ldots, \theta_m\}$.

Before we can assume the chain $\{\theta_0, \theta_1, \ldots, \theta_m\}$ is a series of samples from the posterior distribution $\pi(\theta | Y)$ we must evaluate it for convergence [45, 46]. Convergence diagnostics are important because MCMC algorithms may produce bad posterior estimates if the proposal distribution q is poorly chosen. Typically, proposal distributions are chosen to be symmetric to simplify Eq. (2.5) and also unimodal and symmetric so that proposals are more likely to be similar to the previously accepted value in the chain. However, choosing a good proposal distribution often requires special "tuning" of the parameters that govern the distribution's shape. For example, if the tuning parameter σ in the Gaussian proposal q is small, the chain is likely to get "stuck" in low-probability regions of the posterior. In low-probability regions, the candidate θ^* is unlikely to produce simulated data X close to the observed data Y. However, because σ is small, and the proposal distribution is centered on a low-probability value θ_{t-1}, choosing a proposal θ^* from a distant, high-probability region is very small. In this situation, the probability of the chain moving out of the low-probability region becomes effectively zero. The more often a chain gets stuck, the more likely the samples of the chain are to be highly dependent on one another (i.e., highly correlated). In general, this is an undesirable result because it can impact the resolution of the estimated posterior distribution. One simple remedy for highly dependent chains is a technique known as thinning. Thinning is a procedure where only a subset of the chain consisting of equally spaced samples is retained as a sample from the posterior. For instance, we might decide to keep every third value from $\{\theta_0, \theta_1, \ldots, \theta_m\}$, reducing the length of our chain substantially. Thinning will reduce the degree to which each new sample is dependent on the one before it, but it will also require that we generate much longer chains.

While all MCMC chains are in danger of getting stuck, the ABC MCMC algorithm is particularly susceptible to this because not only must θ^* meet the acceptance probability of the standard Metropolis-Hastings sampler, it must also generate data that are sufficiently close to the observed data. Therefore, the rejection rate of ABC MCMC can be extraordinarily high, requiring an extreme number of computing cycles for even relatively simple problems. To make things worse, MCMC chains cannot be parallelized easily because the state of the chain on iteration t is always dependent on the value of the chain at iteration $t - 1$. This means that we can only perform one iteration at a time. So while ABC MCMC algorithms are easy to program and apply to simple modeling problems, more advanced algorithms are better equipped to handle more complex problems.

2.2.2 Algorithms Using Particle Filtering

Particle filtering is a technique that mitigates, to some extent, the limitation on parallel processing inherent in the MCMC approach. There are many types of particle filtering algorithms, and in this section, we discuss three: partial rejection control (PRC), population Monte Carlo (PMC), and sequential Monte Carlo (SMC). Instead of generating a chain of samples, particle filters start by establishing a large pool of candidate proposals that iteratively converge to a sample from the true posterior distribution. The individual elements in the pool are called particles. At each stage of the algorithm, the particles are perturbed, evaluated, and filtered: a process that brings the pool of proposals closer and closer to a large sample drawn from the desired posterior.

We begin by generating a pool of N candidate values for θ. One approach for initializing the pool is to simply choose a large sample from the prior distribution $\pi(\theta)$. Each particle in the pool carries with it a weight that quantifies the particle's "importance" in the estimation of the posterior distribution. For the first iteration, all the particles are of equal importance and have weight equal to $1/N$. Particles are sampled from the pool with probability proportional to their weights to form the pool in the next iteration. In this next iteration, the particles are perturbed slightly by adding random noise to their location, and the weights assigned to the new particles change according to where the particles are located in the posterior distribution: particles in higher-density regions of the posterior are given more weight, whereas particles in low-density regions are given less weight. This weighting dynamic leads to particles located in low-density regions of the posterior to be sampled less often and occasionally "die out," removing them from the pool entirely. By contrast, particles in high-density regions tend to survive and "give birth" to new particles through the perturbation process. In this way, the particles become concentrated in the region of the true posterior and their relative frequencies in the pool approximate those of a sample from the true posterior.

The process of perturbing and filtering the particles requires a "transition kernel." A kernel function $K(x)$ is defined as a symmetric, non-negative function that integrates to one. That is,

$$K(x) > 0 \text{ for all } x \in (-\infty, \infty),$$

$$K(-x) = K(x), \text{ and}$$

$$\int_{-\infty}^{\infty} K(x)dx = 1. \tag{2.6}$$

The transition kernel used in sequential Monte Carlo sampling serves the same purpose as the proposal distribution in the MCMC algorithm discussed previously, although it is a more general concept. To specify the transition kernel, we need to choose the distribution of a random variable η that will be added to each particle to move it around in the parameter space. For example, if a particle θ^* is sampled from the pool and perturbed by adding a Gaussian deviate $\eta \sim \mathcal{N}(0, \sigma)$ to it, then the new proposed value for θ is $\theta^{**} = \theta^* + \eta$. The transition kernel then describes the distribution for θ^{**} given θ^*: a Gaussian distribution with mean θ^* and standard deviation σ.

Figure 2.2 shows a generic ABC algorithm that uses particle filtering to estimate the posterior distribution of some parameter θ. As in most particle filtering algorithms, there are two stages: an "initialization" stage (i.e., Lines 2–11), and an "evolution" stage (i.e., Lines 12–24). In the initialization stage, particles are first established by sampling them from the prior. Once accepted, these particles form the basis of the first pool, but are assigned equal weights for use in the evolution stage. During the evolution stage, new proposals are selected by first sampling from the previous pool of particles (Line 16) and then adding some random noise around this existing particle (Line 17). Beyond this proposal generation scheme, particle filtering algorithms proceed similar to basic ABC algorithms by generating data

---

1: Given data and model $Y \sim \text{Model}(\theta)$, tolerance thresholds $\varepsilon_{1:T}$, and prior distribution $\pi(\theta)$:
2: At iteration $t = 1$,
3: **for** $1 \leq i \leq N$ **do**
4:     **while** $\rho(X, Y) > \varepsilon_1$ **do**
5:         Sample θ^* from the prior: $\theta^* \sim \pi(\theta)$
6:         Generate data X from θ^*: $X \sim \text{Model}(\theta^*)$
7:         Calculate discrepancy $\rho(X, Y)$
8:     **end while**
9:     Set $\theta_{i,1} \leftarrow \theta^*$
10:    Set $w_{i,1} \leftarrow 1/N$
11: **end for**
12: At iteration $t > 1$,
13: **for** $2 \leq t \leq T$ **do**
14:    **for** $1 \leq i \leq N$ **do**
15:       **while** $\rho(X, Y) > \varepsilon_t$ **do**
16:          Sample θ^* from the previous pool: $\theta^* \sim \theta_{1:N,t-1}$ with probabilities $w_{1:N,t-1}$.
17:          Perturb θ^* by sampling $\theta^{**} \sim N(\theta^*, \sigma)$
18:          Generate data X from θ^{**}: $X \sim \text{Model}(\theta^{**})$
19:          Calculate discrepancy $\rho(X, Y)$
20:       **end while**
21:       Set $\theta_{i,t} \leftarrow \theta^{**}$
22:       Calculate new weights $w_{i,t}$
23:    **end for**
24: **end for**

Fig. 2.2 A generic ABC algorithm that uses particle filtering to estimate the posterior distribution of a parameter θ given data Y

and evaluating its closeness to the observed data Y. Once a particle is accepted, a new weight is assigned to the particle depending on a number of features (Line 22), which we discuss below. At this stage, algorithms such as the ABC PRC and ABC PMC algorithms diverge as they specify different weighting strategies for the particles.

While Fig. 2.2 specifies that particles be perturbed according to a normal distribution (Line 17), some algorithms also require that we specify a transition kernel that takes us back to θ^* from θ^{**}. If the distribution of θ^{**} given θ^* is a "forward" transition kernel, then the distribution of θ^* given θ^{**} is a "backward" transition kernel. If the forward transition kernel is Gaussian as we just described, then, because $\theta^* = \theta^{**} - \eta$, one obvious choice for the backward transition kernel is again a Gaussian distribution with mean θ^{**} and standard deviation σ. In general, the forward and backward kernels need not be symmetric or equal as in this example; in practice, however, they frequently are [47]. The optimal choice for the backward kernel can be difficult to determine [48], and while symmetric kernels greatly simplify the algorithm they can be a poor choice when the posterior is skewed [49].

We now present three sequential Monte Carlo sampling algorithms adapted for ABC. As we will see below, the different sequential Monte Carlo algorithms can be distinguished by how sampling weights are assigned to the particles in the pool across iterations and in the transition kernels they use to perturb the particles. These algorithms are partial rejection control, population Monte Carlo, and sequential Monte Carlo.

2.2.2.1 Partial Rejection Control

The ABC partial rejection control (ABC PRC) algorithm was the first ABC algorithm to use a particle filter [47]. In this algorithm, we must choose both a forward and a backward transition kernel. Using similar notation as above, we denote the forward kernel as $q_f(\theta^{**}|\theta^*)$, and the backward kernel as $q_b(\theta^*|\theta^{**})$. The forward kernel $q_f(\theta^{**}|\theta^*)$ is used to perturb the particle θ^* to θ^{**}, and then we simulate data X using θ^{**} and compare X to the observed data Y by computing the distance function $\rho(X, Y)$. As in the rejection algorithm above, if the particle θ^{**} passes inspection (if $\rho(X, Y)$ is less than some ϵ), then we keep it and assign it a weight that will determine the probability of sampling it on subsequent iterations. If the particle does not pass inspection (if $\rho(X, Y) > \epsilon$), it is discarded, and the process is repeated until we obtain a particle that does pass inspection. The weight w given to the new particle θ^{**} will be

$$w = \frac{\pi(\theta^{**})q_b(\theta^*|\theta^{**})}{\pi(\theta^*)q_f(\theta^{**}|\theta^*)}.$$

This process is repeated until the pool consists of N new particles, each satisfying the requirement that $\rho(X, Y) \leq \epsilon$.

If we stop now, after recreating the pool once, then ABC PRC is equivalent to the simple ABC rejection sampler. However, the major advantage of using the ABC

PRC algorithm is in the gradual filtering process. So, when using ABC PRC, we will typically repeat the process several times. On each iteration, we sample particles with probabilities based on the weights they were assigned in the previous iteration. These weights allow us to discard particles from the pool in low-probability regions (particles said to be "performing poorly") and increase the number of particles in high-probability regions, finally resulting in a sample of particles that represent a sample from the desired estimate of the posterior $\pi(\theta|\rho(X, Y) \leq \epsilon)$.

Another feature of the ABC PRC algorithm is that the values of ϵ can change over iterations. In effect, the filtering process is augmented by making ϵ more stringent over the iterations. On the first iteration, we might adopt a very liberal value for ϵ, call it ϵ_1, such that the resulting pool of particles is only slightly different from the prior distribution from which they were initially sampled. However, on the second iteration, we might choose a new value for ϵ, call it ϵ_2 such that $\epsilon_2 < \epsilon_1$. After a new pool of particles has been obtained, the shape of its frequency distribution will be somewhere between the prior distribution and the desired posterior distribution. In fact, if we can generate a pool such that $\epsilon_t = 0$ and the summary statistics $S(\cdot)$ that define the function $\rho(X, Y)$ are sufficient, then the ABC PMC algorithm produces exact posteriors [29]. However, for continuous measures, because the probability that $\rho(X, Y)$ equals $\epsilon = 0$ is zero, the quality of the approximation will depend on the value of ϵ_T on our final iteration T. One reasonable strategy, therefore, is to set ϵ_1 to a large value and slowly decrease it until the computation time grows impractically large.

This weighting scheme solves several of the problems of ABC MCMC, including the problem of a chain getting stuck in a low-probability region. However, the efficiency of the sampler relies heavily on the choices of the two kernels $q_f(\theta^{**}|\theta^*)$ and $q_b(\theta^*|\theta^{**})$, the sequence of ϵ values, and the prior $\pi(\theta)$. As an example, if we specified a completely uninformative prior (i.e., a flat prior) so that $\pi(\theta)$ is effectively a constant, and we further specify that $q_b = q_f$, then the particle weights w will not change over iterations, and the algorithm reduces to the rejection-based ABC algorithm. Such a scenario, as we argued before, is extremely inefficient and is unlikely to produce good estimates of the posterior distribution. A final consideration is that the ABC PRC produces biased estimates of the posterior [50]. The net effect of this bias is that the distribution defined by the pool of particles and their weights does not necessarily converge to the true posterior. The next algorithm we discuss corrects for this bias using a population Monte Carlo sampling scheme.

2.2.2.2 Population Monte Carlo Sampling

ABC population Monte Carlo sampling (ABC PMC) avoids the biased weighting scheme used in the ABC PRC algorithm [50]. As in the ABC PRC algorithm, the ABC PMC algorithm permits the value of ϵ to change over iterations, allowing particles to move to high-density regions of the desired posterior. While the ABC PRC algorithm requires both forward and backward transition kernels, the ABC PMC algorithm uses a single adaptive transition kernel $q(\theta^{**}|\theta^*)$ whose perturbation variance depends on the distribution of accepted particles in the previous iteration. Due to this dependency, the algorithm is broken into two "stages."

At the first stage, particles are simply sampled from the prior distribution and evaluated based on some ϵ_1 for the first iteration. At this stage, the weights for each corresponding particle are assigned equal weights. Specifically, letting $w_{i,t}$ denote the weight assigned to the ith particle on the tth iteration. Following iteration $t = 1$,

$$w_{i,1} = \frac{1}{N} \ \forall \ i \in \{1, 2, \ldots N\}.$$

The purpose of the initialization stage is not to set weights; instead, the purpose is to gain a sense of the distribution of particles that were accepted. To gain a sense of the spread of these initialized particles, we can simply compute the variance of the particles. Letting $\theta_{i,t}$ denote the ith particle on iteration t, (i.e., with corresponding weight $w_{i,t}$), following the initialization stage we can compute

$$\sigma_t^2 = 2 \frac{1}{N} \sum_{i=1}^{N} \left(\theta_{i,t} - \sum_{j=1}^{N} \theta_{j,t}/N \right)^2 = 2\mathrm{Var}(\theta_{1:N,t}), \tag{2.7}$$

where $t = 1$ at this first stage.

The most novel aspect of the ABC PMC algorithm is the way in which the transition kernel adapts according to the distribution of particles. This feature allows the algorithm to automatically tune itself to gradual changes in the tolerance threshold ϵ_t over iterations. At the second stage (i.e., $t = 2$), we can evolve the pool of particles according to the variance calculated at the previous step. Similar to the ABC PRC algorithm above, we first sample a θ^* from the pervious pool of particles such that $\theta^* \sim \theta_{1:N,t-1}$, with probabilities $w_{1:N,t-1}$. We then introduce some random perturbation noise to the selected particle according to the adaptive transition kernel. For example, if we chose a normal distribution as the transition kernel, the candidate parameter would be generated by sampling $\theta^{**} \sim N(\theta^*, \sigma_{t-1}^2)$. We can then evaluate the proposal θ^{**} in the usual, rejection-based way by comparing its corresponding simulated data to the observed data. Because the particles have been initialized and the transition kernel is set, we can assign weights to the particles on the basis of their density in the posterior distribution relative to the transition kernel. Specifically, the weight assigned to the particle $\theta_{i,t-1}$ is

$$w_{i,t} = \frac{\pi(\theta_{i,t})}{\sum_{j=1}^{N} w_{j,t-1} \ q \left(\theta_{j,t-1} | \theta_{i,t}, \sigma_{t-1} \right)}. \tag{2.8}$$

Following the weight calculation, we can calculate the variance of the particles on this iteration by evaluating Eq. (2.7).

A major concern with any sampling scheme is the speed with which posterior estimates can be obtained. Assuming a fixed model simulation time, the speed of obtaining samples is dictated by the particle acceptance rate, or the probability of accepting a proposal. When proposal distributions or transition kernels are poorly specified, acceptance rates are often very low, which results in a tremendous amount of computation time wasted on evaluating proposals that have no chance of being selected.

With this concern in mind, the ABC PMC weighting scheme optimizes the acceptance probability by setting the variance of the transition kernel to a value that minimizes the Kullback-Leibler distance, a popular statistic that measures the discrepancy between two density functions [50, 51]. In the ABC PMC algorithm, the weights are designed to minimize the distance between the desired posterior distribution and the proposal distribution, such that when the Kullback-Leibler distance is minimized, the acceptance probability is maximized [52].

2.2.2.3 Sequential Monte Carlo Sampling

Developed in parallel to the ABC PMC algorithm, the ABC sequential Monte Carlo (ABC SMC) [49] sampling algorithm relies on a particular type of sequential importance sampling [48]. The weights in ABC SMC are very similar to the weights in ABC PMC, except that the kernel $q(\theta^{**}|\theta^*)$ is nonadaptive (its variance does not change over iterations) and not necessarily Gaussian. In this way, the ABC SMC algorithm is a more general algorithm than the ABC PMC algorithm, and as such it is particularly useful in situations when the transition kernel cannot have infinite support (e.g., cannot be Gaussian). This might happen for certain models in which, for example, the parameter θ cannot be negative. As an example, the probability parameter p in the binomial distribution is bounded between zero and one, making a Gaussian transition kernel an inefficient choice.

Given the generic form of the ABC SMC algorithm, Eq. (2.8) calculates the weight for the ith particle on the tth iteration as

$$w_{i,t} = \frac{\pi(\theta_{i,t})}{\sum_{j=1}^{N} w_{j,t-1} \, q(\theta_{j,t-1}|\theta_{i,t})}.$$

Because the transition kernel is nonadaptive, we must rely on manual adjustments to the tolerance threshold parameter ϵ_t to use the algorithm effectively.

2.2.2.4 Summary

The three sequential Monte Carlo algorithms described in this section demonstrate how particle filters can be used to adaptively estimate posterior distributions in a likelihood-free context. However, these algorithms are all based on a strategy whereby a pool of particles metamorphoses from a sample from the prior distribution to, finally, a sample from the target posterior distribution. Each algorithm achieves this by relying on some sort of filtering process—whether it be in the particle weights or the adaptive nature of the transition kernel. Another successful method of estimation relies on a post hoc adjustment to the values of the posterior samples. These methods are quite general in their approach and can in fact be applied to any of the rejection-based ABC algorithms we discuss here. We describe this method next.

2.2.3 Regression Adjustment

As we discussed in Chap. 1, there is often a systematic relationship between a model's parameter values θ and the resulting summary statistics $S(X)$ calculated from simulated data X. As an example, the left panel of Fig. 1.1 illustrates a relationship between an unknown probability parameter θ for a binomial model and a summary statistic $S(X)$, which is the proportion of "successes" in some binomially distributed data. Each point in the plot represents a particular value of a proposal θ^* and the corresponding distance between the data X produced by simulating the model under θ^* and the observed data Y. In this example, there is a clear linear relationship between these two values. Specifically, as the proposed parameter values approach 0.50, the distance between their corresponding summary statistics approaches 0.0. If zero is the optimal value for $\rho(X, Y)$, then the proposed parameter values resulting in $S(X) = 0$ should have much higher density in the posterior distribution for θ than say, the values resulting in $S(X) = 0.05$.

Recall that the rejection-based algorithms assign a simple reject/accept rule to each parameter proposal, and this rule depends on the value of ϵ. While the particle filter methods turn on adaptive changes to ϵ over the iterations, another approach is to consider the joint relationship between the parameter proposals and the simulated data, as in the left panel of Fig. 1.1. As originally proposed by Beaumont et al. [53], if the relationship between θ and $S(X)$ is approximately linear, we can use linear regression techniques to obtain a corrected estimate for θ. This estimate can then be used to adjust the remaining samples from the approximate posterior distribution.

In this section we will denote the set of posterior samples as Θ, a set of summary statistics as $\mathbf{S}(x)$, and the target (i.e., optimal) value for those summary statistics as $\mathbf{S}_0$. Individual components of the summary statistics are represented with a double subscript reflecting the calculation for the ith sample and mth statistic, such as $S_{i,m}(x)$.

A simple model for linearly regressing the summary statistics $\mathbf{S}(x)$ on the obtained posterior samples Θ is

$$\Theta_i = \alpha + (\mathbf{S}_i(x) - \mathbf{S}_0)^T \beta + \zeta_i, \tag{2.9}$$

where the residuals ζ are independent and identically distributed. When $\mathbf{S}_i(x) = \mathbf{S}_0$, we are drawing samples directly from our desired posterior distribution, whose mean we can denote α. The least squares estimates for α and β are

$$\left(\hat{\alpha}, \hat{\beta}\right) = (X^T X)^{-1} X^T \Theta,$$

where X is the matrix of summary statistics augmented with a column of ones:

$$X = \begin{bmatrix} 1 & S_{1,1}(x) & S_{1,2}(x) & \dots & S_{1,M}(x) \\ 1 & S_{2,1}(x) & S_{2,2}(x) & \dots & S_{2,M}(x) \\ \vdots & \vdots & \vdots & \ddots & \vdots \\ 1 & S_{N,1}(x) & S_{N,2}(x) & \dots & S_{N,M}(x) \end{bmatrix}. \tag{2.10}$$

The strategy is then to adjust the set of posterior samples Θ to have mean α while simultaneously correcting for the trend in the relationship between Θ and $\mathbf{S}(x)$. To do this, we calculate the correction

$$\Theta_i^* = \Theta_i - [\mathbf{S}_i(x) - \mathbf{S}_0]^T \hat{\beta}. \tag{2.11}$$

Following this adjustment, the new set of posterior samples Θ^* will form a random sample from an approximation of the desired posterior distribution.

2.2.3.1 Localized Weighting

While Eq. (2.9) does not make distributional assumptions about ζ, it does assume that there is a linear relationship between Θ and $\mathbf{S}(x)$. This is rarely true in practice, but Beaumont et al. [53] argue that it may be true in a localized region around $\mathbf{S}_0$. Thus, we can perform localized linear regression by applying a weighting function to the posterior samples based on their corresponding $\mathbf{S}(x)$ values. To localize the regression problem, we define a kernel function $K(d)$ that weighs the values of $\mathbf{S}(x)$ as a function of their distance d from the desired $\mathbf{S}_0$. This kernel function can take many forms, such as a Gaussian, exponential, or Epanechnikov. We now define the weight matrix W as

$$W = \begin{bmatrix} K\left(||\mathbf{S}_1(x) - \mathbf{S}_0||\right) & 0 & \cdots & 0 \\ 0 & K\left(||\mathbf{S}_2(x) - \mathbf{S}_0||\right) & \cdots & 0 \\ \vdots & \vdots & \ddots & \vdots \\ 0 & 0 & \cdots & K\left(||\mathbf{S}_N(x) - \mathbf{S}_0||\right) \end{bmatrix}, \tag{2.12}$$

where $||\mathbf{S}_i(x) - \mathbf{S}_0|| = \sqrt{\sum_{j=1}^{M} [S_{i,j}(x) - S_{0,j}]^2}$ is the distance between the vectors of summary statistics $\mathbf{S}_0$ and $\mathbf{S}(x)$. In this localized weighting scheme, the new estimates for α and β from Eq. (2.9) become

$$\left(\hat{\alpha}, \hat{\beta}\right) = (X^T W X)^{-1} X^T W \Theta.$$

The kernel function K may be chosen so that values of $\mathbf{S}(x)$ that are outside the region of $\mathbf{S}_0$ are given a weight of zero, excluding them from influencing the estimates for α and β. This may seem a little wasteful; other methods incorporate these samples by changing the model specification in Eq. (2.9) instead. For example, Blum et al. [33] used nonlinear regression techniques to correct for heteroscedascity. Other applications focus on model selection problems by parameterizing a model selection parameter that conveys the probability that a given model is preferred. For these types of models, one can use logistic regression correction on the model selection parameter to improve the quality of the estimated posterior [54].

Regression correction methods are important for eliminating error that arises from using error terms ϵ that are too large. This feature is often exploited by specifying large values of ϵ so that fewer model simulations are wasted, which improves the efficiency of the sampler. However, because the correction is performed following posterior sampling, opportunities for optimizing the transition kernel on the basis of say, the current pool of particles as in the ABC PMC algorithm, are lost. Furthermore, it is not immediately obvious how one would perform the regression correction in contexts where Gibbs sampling is necessary, as it is for models with many parameters or hierarchical construction. Fortunately, another class of algorithms, called the kernel-based ABC approach, directly incorporates localized weighting in the sampling algorithm itself, and so there is no need for a post-hoc correction. We turn to this approach now.

2.3   Kernel-Based ABC

Rejection-based ABC works well when the model of interest closely matches the observed data; however, it can sometimes be inefficient because of its inability to provide anything but crude evaluations of a proposal's "fitness." In other words, for a given proposal θ^*, if the model produces data such that $\rho(X, Y) > \epsilon$, resulting in a rejection of that proposal, then the density of the posterior $\pi(\theta \mid Y)$ will not just be small but will be zero at the location θ^*.

While we might be tempted to think that zero density where proposals result in $\rho(X, Y) > \epsilon$ is not a bad thing, consider two different proposals θ_1^* and θ_2^*. Suppose θ_1^* is in a merely low-density region of the posterior distribution whereas θ_2^* is far from the posterior distribution (having practically zero density). As a result of the variability in the generation of X_1 for the proposal θ_1^*, it is possible that $\rho(X_1, Y) > \epsilon$. When this occurs, both θ_1^* and θ_2^* will be rejected, resulting in an estimated density of zero at both locations. The problem, then, is the following: if θ_1^* is much closer to the desired posterior than θ_2^*, such that the condition $\rho(X_1, Y) \leq \epsilon$ is sometimes satisfied whereas $\rho(X_2, Y) \leq \epsilon$ is never satisfied, shouldn't this probabilistic information be contained in the weights associated with the proposals θ_1^* and θ_2^*?

Such a weighting procedure has only recently been embedded in ABC samplers [33, 53, 55, 56]. We refer to the class of algorithms that use a continuous weighing procedure as "kernel-based ABC." At their core, kernel-based ABC algorithms take advantage of the localized regression correction techniques discussed above, but they apply localized weighting during the estimation procedure rather than to the samples that have been obtained. To do this, kernel-based algorithms use the idea of model misspecification to apply a weighting procedure to all proposals, regardless of their proximity to the posterior distribution. Weights are computed by assuming that the data Y is a realization of a model simulation under the best possible parameter values $\hat{\theta}$ plus some random error (ζ). This means that

$$Y = \text{Model}(\hat{\theta}) + \zeta, \tag{2.13}$$

where ζ follows a distribution with density $\psi(\delta)$ governed by the tuning parameter δ which determines the variance of ζ.

By assuming that the error ζ is continuous, we can use it to evaluate a particle's fitness. For example, we could choose a Gaussian kernel $\psi(\cdot|\delta)$ function centered at zero and with standard deviation equal to δ. Once selected, the kernel function is used to weight the proposal according to how closely the simulated data matches the observed data $\psi(\rho(X, Y) \mid \delta)$. Because ψ is symmetric, unimodal, and centered at zero (see Eq. (2.6)), and assuming our discrepancy function $\rho(X, Y)$ is chosen appropriately (e.g., a Euclidean metric), then as $\rho(X, Y)$ moves away from zero, $\psi(\rho(X, Y)|\delta)$ will decrease so that a lower weight is assigned to θ. The result of this localized weighting is that proposals θ^* located further from the posterior $\pi(\theta| Y)$ are penalized more heavily because $\psi(\rho(X, Y)|\delta)$ will tend to be smaller, on average. This weighting scheme results in numerical approximation of the equation

$$\pi(\theta| Y) \propto \int_{x \in \mathcal{X}} \pi(\theta) \, \text{Model}(x|\theta) \, \psi(\rho(X, Y)|\delta) \, dx, \tag{2.14}$$

where $\mathcal{X}$ is the range of possible data patterns generated by $\text{Model}(x|\theta)$.

Although the kernel-based ABC algorithms are considerably more efficient than rejection-based ABC algorithms, they still rely on the specification of the standard deviation δ. The standard deviation δ is the kernel-based analog of the error term ϵ in the rejection-based approaches discussed above and carries with it similar interpretations and statistical properties. Specifically, as δ becomes smaller and approaches 0, the approximation becomes more and more accurate. At the same time it becomes increasingly more difficult to sample from the posterior distribution in Eq. (2.14) because fewer and fewer samples will produce satisfactory simulated data X. From a sampling perspective, this creates chains and particle filters that get stuck in low-probability regions of the parameter space, and this in turn requires that more samples be taken (resulting in longer computation times) to obtain good estimates of the posterior distribution $\pi(\theta| Y)$.

2.3.1   Kernel-Based MCMC

As we discussed in the introduction to this chapter, we can choose different methods for generating parameter proposals. One method for instantiating the kernel-based approach is to embed the weighting strategy within a conventional MCMC algorithm, similar to the use of rejection-based methods in the MCMC algorithms discussed above [55]. Figure 2.3 shows a generic kernel-based ABC MCMC algorithm for sampling from the posterior distribution of the parameter θ. First, on Line 2, the chain is initialized to some starting value θ_0 that we know has a nonzero density in the posterior. On iteration 2, we then generate a proposal parameter θ^* by adding random noise to the current state of the chain. One way to add noise, as illustrated on Line 4, is to take a sample from a normal distribution centered at the current state of the chain and having standard deviation σ. From here, Lines 5 and 6 are identical to the rejection-based ABC algorithms above. Next, we must determine whether the chain should move to the location of the

---

1: Given data and model $Y \sim \text{Model}(\theta)$, error term δ, and prior distribution $\pi(\theta)$:
2: Initialize $\theta_1 = \theta_0$.
3: **for** $2 \leq i \leq N$ **do**
4:      Sample θ^* by perturbing it from the chain's location: $\theta^* \sim \mathcal{N}(\theta_{i-1}, \sigma)$
5:      Generate data X from θ^*: $X \sim \text{Model}(\theta^*)$
6:      Calculate discrepancy $\rho(X, Y)$
7:      Sample $p^* \sim U(0, 1)$, and calculate α (Equation 2.14)
8:      **if** $p^* < \alpha$ **then**
9:          Store $\theta_i \leftarrow \theta^*$
10:     **else**
11:         Store $\theta_i \leftarrow \theta_{i-1}$
12:     **end if**
13: **end for**

Fig. 2.3 A kernel-based ABC MCMC algorithm to estimate the posterior distribution of a parameter θ given data Y. $U(0, 1)$ is the continuous uniform distribution over the interval $(0, 1)$

new proposal parameter or stay where it is. Here, we use the Metropolis-Hastings rule with a special twist for application to the likelihood-free context. Letting X^* be the simulated data associated with the proposal θ^* and X_t be the simulated data associated with the state of the chain on the tth iteration, we can write the Metropolis-Hastings probability as

$$\alpha = \min\left(1, \frac{\pi(\theta^*)\psi(\rho(X^*, Y)|\delta)q(\theta_t|\theta^*)}{\pi(\theta_t)\psi(\rho(X_t, Y)|\delta)q(\theta^*|\theta_t)}\right). \tag{2.15}$$

Equation (2.15) is similar to what was shown in Eq. (2.5), with the additional weight term $\psi(\rho(X, Y)|\delta)$, which effectively approximates the likelihood function. Once α has been calculated, we accept the new proposal θ^* with probability α, otherwise we reject θ^*, and the chain will stay put. To implement a probabilistic acceptance rule, we can simply draw a value p^* randomly from the unit interval such that $p^* \sim U(0, 1)$. If α happens to be larger than the drawn p^*, we keep the proposal as shown on Lines 8–10 in Fig. 2.3.

In Wilkinson [55], the error standard deviation δ was constrained to be the same across all iterations of the algorithm, but this is not the only choice. For example, in parallel to a gradually changing ϵ term in the rejection-based algorithms above, so too can δ change over iterations. While MCMC sampling worked well in the examples presented by Wilkinson [55], other problems are more difficult for MCMC algorithms to handle, such as when the number of dimensions is high or the parameters exhibit strong parameter-to-parameter correlations [57]. In these cases, picking one δ term across all iterations may lead to undesirable sampling behavior, and as a result, poor estimates of the posterior distribution. In addition, the seemingly innocuous choice of setting the tuning parameter σ in Fig. 2.3 becomes difficult in these scenarios, and poor choices can lead to poor efficiency in the sampling algorithm. In the next section, we discuss how a different proposal scheme can help in generating proposal parameters more efficiently.

2.3.2 ABC with Differential Evolution

Another kernel-based algorithm is the Approximate Bayesian Computation with Differential Evolution (ABCDE) algorithm, which combines the traditional ABC framework with differential evolution (DE) to generate proposals [57–60]. DE is an extremely efficient population-based genetic algorithm for performing optimization and, for our purposes, exploring high-dimensional parameter spaces [61]. As we have discussed throughout this chapter, most proposal-generating schemes rely on random perturbations of each member of a class of either particles or chains to drive convergence to the desired posterior distribution, and there is little to no communication between the chains or particles. By contrast, DE creates a self-organizing population of members by evolving each member based on a weighted difference between other members of the population, similar in spirit to particle swarm optimization [62–64]. In other words, the members of the population of proposals communicate valuable information about the shape and location of the desired posterior distribution, a process that results in a dramatically more efficient estimation method.

The ABCDE algorithm has three steps: crossover, mutation, and migration [59]. Figure 2.4 shows the basic structure of the ABCDE algorithm in pseudocode. We begin by selecting a pool of P particles and dividing this pool evenly into K groups of size $G = P/K$. The particles within each group are not independent, as in the particle filtering methods we presented earlier: the groups are pools of Markov chains that interact, forming an entire system similar to population Monte Carlo samplers.

After the groups are formed, the particles within each group are initialized by sampling values from the prior $\pi(\theta)$, which is represented in Lines 1–2 of Fig. 2.4. After all particles are initialized, the next step in the algorithm is to decide whether or not a migration step should be performed with probability α. The migration step is the distributed genetic algorithm method of diversifying each of the groups, and it is represented in Fig. 2.4 by Lines 4–6. To perform the migration step, we first determine the number of groups η that will be involved in the step. To do this, we sample η groups uniformly from the set $\mathscr{K} = \{1, 2, \ldots, K\}$ without replacement, forming the group set $\mathscr{G} = \{\mathscr{G}_1, \mathscr{G}_2, \ldots, \mathscr{G}_\eta\}$. Next, for each group in $\mathscr{G}$ we sample a single particle θ^* with probabilities based on the inverse of the current weight corresponding to θ^*. Finally, we swap the particles from each of the sets in a cyclical fashion so that

$$\{\theta^*_{\mathscr{G}_1}, \theta^*_{\mathscr{G}_2}, \ldots, \theta^*_{\mathscr{G}_{\eta-1}} \; \theta^*_{\mathscr{G}_\eta}\} \rightarrow \{\theta^*_{\mathscr{G}_\eta}, \theta^*_{\mathscr{G}_1}, \ldots, \theta^*_{\mathscr{G}_{\eta-2}}, \theta^*_{\mathscr{G}_{\eta-1}}\}.$$

```
 1: Initialize θ₁:K,1:G,1 by sampling from the prior:
 2: θ₁:K,1:G,1 ~ π(θ)
 3: for 2 ≤ t ≤ T do
 4:     if p₁* < α then
 5:         θ₁:K,1:G,t ← Migrate(θ₁:K,1:G,t−1)
 6:     end if
 7:     for 1 ≤ k ≤ K do
 8:         if p₂* < β then
 9:             θk,1:G,t ← Mutate(θk,1:G,t−1)
10:         else
11:             θk,1:G,t ← Crossover(θk,1:G,t−1)
12:         end if
13:     end for
14: end for
```

Fig. 2.4 The ABCDE algorithm for estimating the posterior distribution of θ.

Unlike the mutation and crossover steps, we recommend that the migration step be deterministic, and so it will not rely on the Metropolis-Hastings probability in Eq. (2.5). However, probabilistic rules can be adopted so that each swap is determined by first adding a small amount of random noise as proposed by Hu and Tsui [59].

After the migration step, the ABCDE algorithm updates the particles by means of a crossover step (the core DE particle update mechanism, described below) performed with probability β or by means of a mutation step with probability $1 - \beta$. The mutation step occurs in a way that is similar to other MCMC or particle filtering steps where particles are perturbed from their location by means of a generic transition kernel. Once the particles have been perturbed, they are given weights according to the Metropolis-Hastings probability, given in Eq. (2.5). The decision to update via the crossover or the mutation steps is performed for each of the K groups, and is represented in Fig. 2.4 by Lines 7–13.

The most powerful mechanism in ABCDE is the way that it generates efficient proposals via the crossover step [60, 61]. Essentially, the crossover step allows particles that are performing well (i.e., have a high posterior density) to be selected to guide other poorly performing particles to higher-density regions. In the ABCDE algorithm, the crossover step is used to update the individual groups of particles, allowing each sub-population to evolve largely independently. The division of the pool of particles into sub-populations prevents the algorithm from falling into local minima [65].

Figure 2.5 (right panel) illustrates the crossover step for a single group of particles. Three particles from the group are selected at random (without replacement):

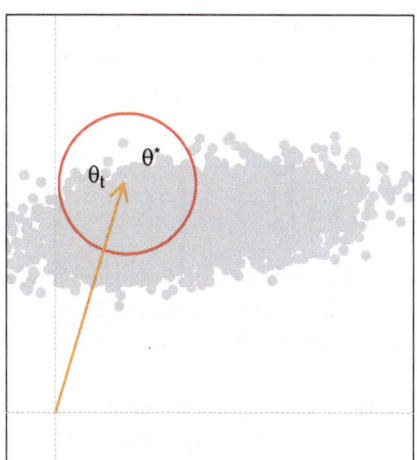 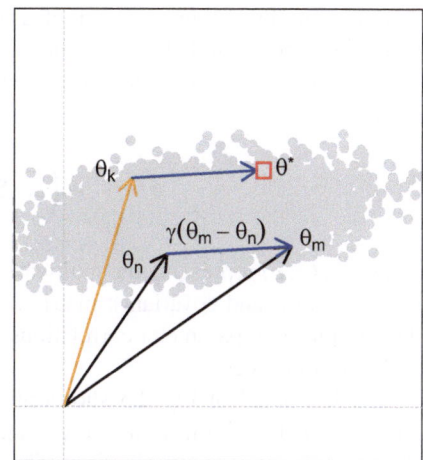

Fig. 2.5 Graphical representation of the perturbation method assuming an independent bivariate normal transition kernel (left panel) used in conventional MCMC and the crossover method (right panel) used in ABCDE

k, m, and n, who have states θ_k, θ_m, and θ_n, respectively. A new proposal (θ^*) is computed as

$$\theta^* = \theta_k + \gamma(\theta_m - \theta_n) + \epsilon: \tag{2.16}$$

the current state of particle k plus some proportion of the scaled difference of the states of particles m and n, plus a small perturbation ϵ. Figure 2.5 shows the difference between θ_m and θ_n as a vector, the blue line, that determines the direction in which the proposal (θ^*) should move relative to θ_k. When the direction of movement has been determined, the tuning parameter $\gamma > 0$ determines the magnitude of the move. Larger values for γ produce proposals that move more aggressively around the parameter space, which can help the algorithm to explore more quickly, but can also result in many more proposals being rejected. The random perturbation ϵ avoids problems of degeneracy. The influence of ϵ on θ^* is illustrated by the red region in the right panel of Fig. 2.5.

One particularly advantageous feature of the crossover step is the way that it handles posterior distributions with highly correlated parameters. When a model has highly correlated parameters, the safest choice is often to ignore the correlation and select a separate tuning parameter (e.g., a variance parameter in a multivariate Gaussian) for each dimension in the parameter space. In psychology, this is often done for "blocked" sampling, where the parameter space is divided into groups of parameters that can be easily sampled simultaneously [9–11,45]. Depending on how highly correlated the parameters in the model are, ignoring their correlation can lead to very poor sampling behavior.

For example, suppose our target distribution contains two parameters that are moderately correlated. If we do not know how correlated these parameters are, we might naïvely choose an independent multivariate normal proposal distribution such that

$$\theta^* \sim \mathcal{N}_2 \left(\theta_t, \Sigma = \begin{bmatrix} \sigma_{11}^2 & 0 \\ 0 & \sigma_{22}^2 \end{bmatrix} \right), \tag{2.17}$$

where $\mathcal{N}_p(a, b)$ represents the multivariate normal distribution of dimension p with mean vector a and covariance matrix b. Because the covariance matrix Σ ignores the possibility of parameter correlations, this proposal distribution can result in very high rejection rates.

The left panel of Fig. 2.5 shows an example of a sampling problem using an independent transition kernel (e.g., Eq. (2.17)) when estimating a target density, shown as the gray cloud of points. The vector shows the current state of the chain θ_t and the proposal region for θ^*, shown as a red circle. If we assume that the parameters are uncorrelated, we will generate many proposals θ^* that are not in the target density, a situation that is illustrated by the proportion of white area inside the circle. Proposals generated in this area will almost certainly be rejected.

Referring back to Fig. 2.4, in practice we recommend that the two tuning parameters α and β be small (e.g., $\alpha = \beta = 0.10$), so that most of the particle updates occur via the DE crossover step, and migration only occurs a few times throughout the simulation. Migration will have the effect of pulling so-called "outlier" particles back into the group while keeping the diversity of the individual groups high enough so that the parameter space can be traversed more effectively.

The ABCDE algorithm's ability to estimate high-dimensional posterior distributions and automatically tune its proposal-generating mechanism to maximize efficiency has led to a number of successful applications of the algorithm to interesting problems in psychology. We will discuss a few of these applications in Chaps. 4 and 5.

2.4 General Methods

The final suite of non-hierarchical techniques are substantially different from either the rejection-based or kernel-based algorithms in that neither require tolerance thresholds nor do they require assumptions about the distributions of summary statistics. However, these workarounds do come with the potentially high price of additional computational complexity. We now discuss each of these methods, synthetic likelihood and probability density approximation, in turn.

2.4.1   Synthetic Likelihood

The synthetic likelihood approach goes one step beyond the kernel-based approaches discussed in the previous section. Wood [43] proposed the synthetic likelihood algorithm as a method for likelihood-free parameter estimation that, unlike previous likelihood-free algorithms, does not require the use of error terms such as ϵ in rejection-based algorithms or δ in kernel-based algorithms that can sometimes produce inaccurate posteriors. Because it does not use a tolerance threshold (as in the other algorithms we have discussed so far), nor does it require a specification of a kernel function, we consider it a "general" method because its distributional assumptions over the selected summary statistics are well founded in statistical theory.

To implement the synthetic likelihood algorithm, we first generate a proposal value θ^* and simulate J new data sets of the same size N and design as the observed data so that $X = \{X_1, X_2, \ldots, X_J\}$, where $X_j = \{X_{j,1}, X_{j,2}, \ldots, X_{j,N}\}$. For the jth simulated data set X_j, we then compute a vector of summary statistics $S^{(j)}(X_j) = \{S_1^{(j)}(X_j), S_2^{(j)}(X_j), \ldots, S_M^{(j)}(X_j)\}$. The summary statistics across the J simulated data sets are then used to compute a mean vector

$$\hat{\mu}_\theta = \frac{1}{J} \sum_{j=1}^{J} S^{(j)}(X_j),$$

where the covariance matrix $\hat{\Sigma}_\theta = QQ^T / (J - 1)$, and

$$Q = [S^{(1)}(X_1) - \hat{\mu}_\theta, S^{(2)}(X_2) - \hat{\mu}_\theta, \ldots, S^{(J)}(X_J) - \hat{\mu}_\theta].$$

The same summary statistics are computed for the observed data, which we denote $S(Y)$ (i.e., without the super script index), and we assume they have the parametric form

$$S(Y) \sim \mathcal{N}(\mu_\theta, \Sigma_\theta).$$

Finally, using the normality assumption and the central limit theorem, the log synthetic likelihood function is given by

$$\mathcal{SL}(\theta|Y) = -\frac{1}{2}(S(Y) - \hat{\mu}_\theta)^T \hat{\Sigma}_\theta^{-1} (S(Y) - \hat{\mu}_\theta) - \frac{1}{2} \log |\hat{\Sigma}_\theta|. \tag{2.18}$$

The synthetic likelihood $\mathcal{SL}(\theta|Y)$ now takes the place of the unknown likelihood, and we can sample from the posterior distribution as in the other likelihood-free methods discussed above by combining the synthetic likelihood approximation with the prior distribution:

$$\pi(\theta|Y) \propto \mathcal{SL}(\theta|Y)\pi(\theta).$$

While the synthetic likelihood approach has certain advantages over other likelihood-free algorithms, the disadvantage of using this approach is that it is more computationally costly: We must generate multiple data sets of similar size to the observed data to evaluate just a single proposal of the parameter values. Perhaps a greater disadvantage is the continued reliance on the assumption that the chosen summary statistics are sufficient for the parameters θ.

2.4.2 Probability Density Approximation

The probability density approximation method differs from other likelihood-free algorithms in two substantial ways. First, the PDA method makes no assumption that a set of summary statistics be jointly sufficient for the parameters of interest. Second, the PDA method is a nonparametric approach, and so it does not require any restrictive assumptions about the distribution of the summary statistics $S(\cdot)$ (as required by the synthetic likelihood algorithm).

Again let $Y = \{Y_1, Y_2, \ldots, Y_N\}$ arise from a model so that $Y \sim \text{Model}(\theta)$. We begin by generating a proposal θ^*. The method for generating θ^* can be one of many options, such as MCMC, ABCDE, or particle filtering. We then use θ^* to simulate a set of data $X = \{X_1, X_2, \ldots, X_J\}$ from the assumed model, so that $X \sim \text{Model}(\theta^*)$. Next, we estimate the form of the random distribution of X, which we call the "simulated probability density function" (SPDF) and denote $f(x|X)$. The SPDF can be obtained using any of a number of density estimation techniques, which we discuss below. Using the SPDF, we evaluate the density of the observed data Y under a given θ by the equation

$$\text{Model}(Y_i|\theta) = f(Y_i|X). \tag{2.19}$$

Thus, after evaluating Eq. (2.19), we obtain a density under the assumed model for every point in the data set Y. Because the data are always sufficient to themselves, our density estimation procedure guarantees sufficiency because the summary statistics are computed for each individual observation Y_i.

Following Eq. (1.2), an approximation of the likelihood function is

$$\mathscr{L}(\theta|Y) = \prod_{i=1}^{N} \text{Model}(Y_i|\theta). \tag{2.20}$$

The "pseudo-likelihood" for a given proposal θ^* is obtained by computing $\mathscr{L}(\theta^*|Y)$. We can then approximate the posterior density, up to a constant of proportionality, with

$$\pi(\theta|Y) \propto \mathscr{L}(\theta|Y)\pi(\theta). \tag{2.21}$$

Equation (2.19) forms the necessary approximation used in the evaluation of the likelihood function, and ultimately the posterior distribution. However, the method that we choose for evaluating Eq. (2.19) should take into account the characteristics of the observed measures. Turner & Sederberg [38] discussed three situations: discrete data, continuous data, and mixed data (i.e., containing both continuous and discrete measures). For completeness, we will recreate the discussions of the discrete and continuous versions of PDA below. However, we refer the reader to [38] for a detailed exposition of the PDA method for mixed data, because that version of the algorithm can be viewed as a merger of the discrete and continuous cases below.

2.4.2.1 Discrete Data

Discrete data are common in psychology, and can be found in the form of confidence or rating responses (e.g., a Likert scale), and response frequencies, such as the number of hit and false alarms (and by extension, hit and false alarm rates). For discrete measures, the SPDF $f(x|X)$ is constructed by means of a relative frequency distribution.

We first define a sample space $S = \{s_1, s_2, \ldots, s_n\}$ as the set of all possible outcomes in our experiment. Randomness in the process of model simulation implies that the simulated data X are random variables with their own sample space from which we observe the random data Y. If we define the set of possible simulated outcomes as $\mathscr{X} = \{x_1, x_2, \ldots, x_m\}$, we can define SPDF as

$$f(x|X) = P(s_j \in S : X(s_j) = x_i), \tag{2.22}$$

which restricts the sample space of the simulated data $\mathscr{X}$ to lie in the sample space S of the experiment. Equation (2.22) states that the probability of X equaling a given value of x_i is the number of times that the simulated data equaled the given value x_i, divided by the total number of model simulations. The SPDF is then used in Eqs. (2.19) and (2.20) to evaluate the pseudo-likelihood of the proposal θ^*.

2.4.2.2 Continuous Data

When the data Y have continuous measurements, we cannot use a relative frequency distribution to characterize the random distribution of Y. Instead, we must use the simulated data X to form an approximation of the likelihood by estimating a density function. While there are many ways of estimating a density function, one of the most useful is the nonparametric kernel density estimate [39].

We proceed in the same way as in the discrete case by first generating a proposal θ^* and using the proposal to generate a sequence of observations X from the model, so that $X \sim \text{Model}(\theta^*)$. We then construct a kernel density estimate of the simulated data so that

$$f(x|X) = \frac{1}{hJ} \sum_{j=1}^{J} K\left(\frac{x - X_j}{h}\right). \tag{2.23}$$

The function $K(\cdot)$ is the kernel and h is a smoothing parameter known as the bandwidth. The kernel satisfies the conditions of Eq. (2.6), and so ensures that $f(x|X)$ is a true probability density. The kernel K is usually chosen to be continuously decreasing as its argument moves away from zero, which places decreasing weights on observations X_j further from the point x where the density is being estimated. While the kernel can take many forms, in practice we have only used the Epanechnikov kernel, given by

$$K(x) = \begin{cases} \dfrac{3}{4}\left(1-x^2\right) & \text{if } x \in [-1, 1] \\ 0 & \text{if } x \notin [-1, 1] \end{cases}. \qquad (2.24)$$

The accuracy of a kernel density estimate is measured by the mean integrated squared error (MISE), a measure of divergence between a true and an estimated density function. The Epanechnikov kernel was derived as a solution that minimizes the asymptotic MISE, and so it is optimal in a statistical sense [39, 66]. To select a bandwidth h, we use Silverman's [39] rule of thumb, so that

$$h = 0.9 \min\left(\text{SD}(X), \frac{\text{IQR}(X)}{1.34}\right) n^{-1/5}, \qquad (2.25)$$

where $\text{SD}(X)$ denotes the standard deviation of X, and $\text{IQR}(X)$ denotes the interquartile range. While these choices work well for all of our examples and are the standard methods available in the scientific libraries in R and Python, it is likely that our method of kernel density estimation could be further improved, especially in the case of small samples [67, 68].

After we have constructed the SPDF $f(x|X)$ via kernel density estimation, we can calculate the pseudo-likelihood function by evaluating Eq. (2.20), where $\text{Model}(x|\theta) = f(x|X)$. The posterior density is then determined by Eq. (2.21). Equations (2.19) and (2.20) together show that each data point is used in the evaluation of the likelihood, so there is no compression of the observed data into summary statistics.

2.5   Hierarchical ABC Algorithms

Psychologists are often interested in systematic differences between groups or people. Subject-specific details such as age, demographic factors, or gender may be expected to influence a person's performance on different tasks. One naïve approach to understanding these individual differences is to assume that they manifest as differences in parameters across different people, and so we might estimate model parameters for each person independently. We could then use these parameter estimates as measurement observations in the same way that we might treat the measurement of the dependent variables. That is, we could perform inferential statistical analysis on the estimated parameters to draw conclusions about the influence of the experimental conditions on the underlying data-generating mechanism.

However, another approach is to assume that the individual-level parameters share some commonality, a relationship that is described by "group-level" parameters. One could then *simultaneously* estimate the parameters specific to each person and the parameters that are common to the group in a hierarchical structure. This approach is called hierarchical modeling.

Despite growing in popularity, ABC is difficult to implement for hierarchical models. When the number of parameters is small (a low-dimensional problem), ABC algorithms can be naïvely extended to hierarchical designs by jointly estimating the parameters across the tiers of the hierarchy—individual-level parameters are sampled and rejected at the same time as the group-level parameters. For example, Turner & Van Zandt [44] showed how the parameters of a simple binomial model could be extended and estimated in a hierarchical framework. Such algorithms, which estimate all parameters of the model in a single stage, have been the basis of the chapter so far. However, as the number of parameters increases, the feasibility of using these single-step algorithms for hierarchical models becomes impractical. Instead, we must consider methods to adapt the single-step algorithm to hierarchical problems. Figure 2.6 shows a naïve solution to this problem in which the hyperparameter vector ξ is updated in a single-step together with the parameters θ.

Figure 2.6 breaks the sampling problem into two steps. First, we sample proposal hyperparameters ξ^* from the prior $\pi(\xi)$, and then we sample proposal parameters θ_j^* from the conditional prior $\pi(\theta_j|\xi^*)$. Writing the data as $Y = \{Y_1, Y_2, \ldots, Y_J\}$, so that Y_j represents Subject j's observations, each set of parameters $\{\xi^*, \theta_j^*\}$ must generate data X_j^* satisfying $\rho(X_j^*, Y_j) \leq \epsilon$ (if using a rejection-based approach). If a candidate hyperparameter ξ^* produces θ_j^*s that satisfy the criterion for all $j \in \{1, 2, \ldots, J\}$, then ξ^* and the θ_j^*s have some nonzero density in the approximate joint posterior distribution $\pi(\xi, \theta|\rho(X, Y) \leq \epsilon)$.

This idea has been implemented in the genetics literature to analyze mutation rate variation across specific gene locations [28, 69]. However, as dimensionality increases, as it would, for example, in an experimental design with a large number of people, the naïve approach presented in Fig. 2.6 would be very slow (and even impractical) because of the overwhelmingly higher rejection rate [29]. This is because it may not be possible to find a θ_j^* that produces X_j^* close to Y_j, even if all the other $\theta_{l \neq j}^*$ produced X_l^*s close to their Y_ls. In this case, the proposed ξ^* and all the proposed θ_j^*s must be discarded, and the search for a sample of θ begins again with a new ξ^*. Therefore, this algorithm, while producing accurate estimates of the posterior, is hopelessly inefficient for even moderately complex problems because of the high-dimensional nature of the proposal scheme for the θ_j^*s.

A more reasonable approach would be to partition the parameter space on the basis of how the conditional distributions interact with the likelihood function. This partitioning, as this chapter has suggested, is a computationally demanding aspect of likelihood-free algorithms because it must be approximated numerically through simulation.

```
 1: for 1 ≤ i ≤ N do
 2:     Sample ξ* from the prior: ξ* ~ π(ξ)
 3:     for 1 ≤ j ≤ J do
 4:         Sample θ*_j from the prior:
 5:             θ*_j ~ π(θ | ξ*)
 6:         Generate data X*_j using the model:
 7:             X*_j ~ Model(θ*_j)
 8:     end for
 9:     Determine jump probability α.
10:     Generate p* ~ U(0, 1)
11:     if p* < α then
12:         Store θ_{1:J,i} ← θ*_{1:J}
13:         Store X_{1:J,i} ← X*_{1:J}
14:         Store ξ_i ← ξ*
15:     else
16:         Store θ_{1:J,i} ← θ_{1:J,i-1}
17:         Store X_{1:J,i} ← X_{1:J,i-1}
18:         Store ξ_i ← ξ_{i-1}
19:     end if
20: end for
```

Fig. 2.6 A naïve algorithm for estimating the posterior distribution of θ and ξ for a hierarchical model. $U(0, 1)$ is the continuous uniform distribution over the interval $(0, 1)$

2.5.1 The Gibbs ABC Algorithm

To remedy the computational inefficiencies observed in other hierarchical ABC methods, Turner and Van Zandt [70] proposed the Gibbs ABC algorithm. In this approach, parameters are divided into two groups: individual-level parameters that depend on the likelihood function and group-level parameters that do not. The key to using Gibbs ABC is that the posterior of a set of hyperparameters depends on the data only through the lower-level parameters, which means that approximating the likelihood function is necessary only when updating the individual-level parameters. In other words, we can employ Gibbs sampling at the level of the hyperparameters and bypass the problem of dimensionality and numerical errors that occur in approximating the likelihood altogether.

To implement Gibbs ABC, we first consider the conditional posterior distribution of the individual-level parameters θ, which is

$$\pi(\theta \,|\, Y, \xi) \propto L(\theta \,|\, Y, \xi)\pi(\theta|\xi)$$

$$\propto \prod_{j=1}^{J} L(\theta_j | Y_j) \pi(\theta_j | \xi),$$

given the conditional independence of the θ_js and Y_js. The parameter ξ drops out of the likelihood because ξ has no role in the probability density of the data Y_j. The likelihood Y_j depends on ξ only through the parameters θ_j. We can go one step further and separate the effects that are exclusive to Y_j by noting that, by the independence of the Y_js and θ_js,

$$\pi(\theta_j | Y, \xi) \propto L(\theta_j | Y_j) \pi(\theta_j | \xi). \tag{2.26}$$

Using an approximation $\psi(\cdot | \delta)$ of the likelihood for the jth person, Eq. (2.26) becomes

$$\pi(\theta_j | Y, \xi) \approx \psi(\rho(X_j, Y_j) | \delta) \pi(\theta_j | \xi). \tag{2.27}$$

In this discussion we assume that $\psi(\cdot | \delta)$ is based on the kernel density estimation method, but we could have computed just as well the approximation with rejection-based methods.

The final step is to derive the conditional distribution for the hyperparameters. Noting that $\pi(\xi | Y, \theta) \propto \pi(\theta | \xi) \pi(\xi)$, the joint conditional posterior distribution of the hyperparameters ξ is

$$\pi(\xi | Y, \theta) \propto L(\theta | Y) \pi(\theta | \xi) \pi(\xi)$$

$$\propto \pi(\theta | \xi) \pi(\xi)$$

$$\propto \pi(\xi) \prod_{j=1}^{J} \pi(\theta_j | \xi). \tag{2.28}$$

Because ξ influences the likelihood only through the parameter θ, the joint conditional distribution of $\xi = \{\xi_1, \ldots, \xi_m, \ldots, \xi_M\}$ does not depend on the likelihood; the likelihood is a constant with respect to ξ. This means that any of the algorithms we have discussed up to this point can be used to approximate Eq. (2.27), whereas we can sample ξ from its conditional posterior distribution using standard techniques.

Figure 2.7 shows pseudocode for the Gibbs ABC algorithm. After initializing values for $\xi_{1:M,1}$ and $\theta_{1:J,1:K,1}$ on iteration $i = 1$, on each iteration $i \geq 2$, we perform the following steps. First we draw samples of $\xi_{m,i}$ conditioned on all other parameters in the model, including all other values in the vector ξ. Then we use a Gibbs sampler to obtain values of $\xi_{1:M,i}$ by sampling directly from the posterior $\pi(\xi_m | Y, \theta_{1:J,1:K,i-1}, \xi_{-m,i})$ given by Eq. (2.28).

The Gibbs ABC algorithm is considerably more flexible than other hierarchical ABC algorithms. Specifically, Gibbs ABC can be used as a mixture of likelihood-free and likelihood-informed techniques, depending on which parameters are being

```
 1: Initialize ξ_{m,1} and each θ_{j,k,1}.
 2: for 2 ≤ i ≤ N do
 3:     for 1 ≤ m ≤ M do
 4:         Sample ξ_{m,i} from the conditional posterior:
 5:             ξ_{m,i} ∼ π(ξ_m | θ_{1:J,1:K,i−1}, ξ_{−m,i})
 6:     end for
 7:     for 1 ≤ j ≤ J do
 8:         for 1 ≤ k ≤ K do
 9:             Sample a value θ*_{j,k} from a proposal distribution: θ*_{j,k} ∼ q(θ)
10:             Generate data X*_{j,k} using the model: X*_{j,k} ∼ Model(θ_{j,−k,i}, θ*_{j,k})
11:             Determine jump probability α and sample p* ∼ U(0, 1).
12:             if p* < α then
13:                 Store θ_{j,k,i} ← θ*_{j,k}
14:                 Store X_{j,k,i} ← X*_{j,k}
15:             else
16:                 Store θ_{j,k,i} ← θ_{j,k,i−1}
17:                 Store X_{j,k,i} ← X_{j,k,i−1}
18:             end if
19:         end for
20:     end for
21: end for
```

Fig. 2.7 The Gibbs ABC algorithm to estimate the posterior distributions for ξ and θ

updated. We can use any appropriate sampling method to estimate the posterior distribution of ξ, while approximating the posterior distribution of each θ_j. There is also no reason why we couldn't use different tuning parameters δ for each person. This might be useful when the model is misspecified, such that allowing for large distances for some subjects could improve convergence speed.

As Gibbs sampling does in standard, likelihood-informed Bayesian analysis, the Gibbs ABC algorithm also permits blocked sampling of parameters. Blocking can aid in the convergence of the algorithm, especially in cases where parameters are highly correlated. However, blocking in Gibbs sampling has been shown to provide sampling efficiency that is similar to the DE proposal scheme discussed above when the parameters of a model are highly correlated [57].

2.5.1.1 A Hierarchical Poisson Example

Although Fig. 2.7 presents pseudocode for implementing the Gibbs ABC algorithm, because the algorithm can be complex in more realistic scenarios, we now turn to an illustrative example. For this illustrative example, we consider data arising from a Poisson distribution, where we will model both subject-level and group-level effects in a hierarchical model. As we will see below, the Poisson distribution is a convenient example for illustration as the conditional distributions of the subject- and group-level parameters are tractable, and so we can compare the estimates obtained with the Gibbs ABC algorithm to those obtained with a conventional Gibbs sampler.

The data could come from a variety of experimental paradigms, but suppose we have $J = 4$ subjects who provide a number of "incidents" each week for $T = 100$ weeks. Hence the data Y will consist of 100 incident counts, representing the number of incidents per week, for each of four subjects. To begin building a hierarchical Bayesian model, we first assume that the number of incidents Subject j has during the tth week can be modeled with a Poisson distribution

$$Y_{j,t}|\theta_j \sim \text{Poisson}(\theta_j), \tag{2.29}$$

where θ_j is the incident rate parameter for the jth subject. To extend the model hierarchically, we will assume that the incident rates $\theta = \{\theta_1, \theta_2, \ldots, \theta_J\}$ come from an exponential distribution with rate parameter λ, such that

$$\theta_j|\lambda \sim \text{Exp}(\lambda). \tag{2.30}$$

The parameter λ represents the overall incident rate for the group of subjects, where larger values produce smaller numbers of incidents per week.

To generate simulated data from this model in R, we can first simply choose a value for λ to dictate the rates of the θ_j parameters. From there, we can use the generated θ_j parameters to produce the data Y. The following block of code can be used to generate our simulated data, where J is the number of subjects and T are the number of weeks:

```
J=4           # number of subjects
T=100         # number of time points

lambda=.3                   # value of the hyperparameter
theta=rexp(J,lambda)        # generate parameters for each subject
Y=matrix(NA,J,T)            # data matrix
for(j in 1:J){Y[j,]=rpois(T,theta[j])} # generate subject data
```

If we were to perform inference on the parameters θ and λ in a Bayesian setting, we could use Eqs. (2.29) and (2.30) to specify a hierarchical model. The final step would be to specify a prior on the parameter λ. Assuming we knew nothing at all about these incident rates, we could specify a noninformative prior, such that

$$\lambda \sim \Gamma(0.01, 0.01). \tag{2.31}$$

With this prior, we can work through the math to derive a conditional distribution for the parameter λ, which also turns out to be a gamma distribution of the form:

$$\lambda|\theta, Y \sim \Gamma\left(J + 0.01, \sum_{j=1}^{J}\theta_j + 0.01\right). \tag{2.32}$$

Thus, the conditional distribution for λ has a convenient form from which we can sample directly. Notice that the conditional distribution does not depend on the data Y, as they are conditionally independent, a feature that is exploited in the Gibbs ABC algorithm. We could also derive the conditional distribution for the θ_j parameters, but as our application is intended to be likelihood-free, we are assuming that the Poisson model's likelihood function is unknown, and so we would not be able to derive the conditional distribution.[2] Instead, we must rely on simulating data from the Poisson model and approximating the conditional distribution through Eq. (2.27).

To approximate the model parameters using a likelihood-free technique, we must first choose an algorithm. Consulting Table 2.1, we might choose the rejection-based algorithm combined with Gibbs ABC as the number of parameters to be estimated are few in number. For a rejection-based algorithm, we must choose a set of summary statistics and a tolerance threshold. As in the example above, we might choose a distance function of the form

$$\rho(X_j, Y_j) = \frac{1}{T} \left| \sum_{t=1}^{T} X_{j,t} - \sum_{t=1}^{T} Y_{j,t} \right| \tag{2.33}$$

to compare the jth subject's data to data generated by the model. Note that because we are only updating the parameters θ_j for a single subject at a time, we need only specify a function to compare this subject's data relative to simulated data from the model. As for ϵ, we can choose something small as the number of parameters are small, and we would like a close approximation of the posterior. For this example, let's set $\epsilon = 1 \times 10^{-10}$.

Given these choices, we can now declare variables corresponding to these parameters within R. The following block of code instantiates ϵ (Line 1), sets the number of samples (Line 2), sets the parameters of the prior distribution from Eq. (2.31) (Line 3), and the ρ function from Eq. (2.33) (Line 4):

```
eps=1e-10                  # tolerance threshold
N=1000                     # number of particles
p.lambda=c(.01,.01)        # prior settings (Gamma distribution)
rho=function(x,y) abs(sum(x)-sum(y))/T      # rho function
```

With the data Y generated and the settings of our algorithm specified, we can now run the sampler to estimate the parameters θ and λ. As the sampler is rejection-based at the level of θ, but is drawing samples from the conditional distribution of λ, the next block of code should look similar to the rejection-based algorithm we presented above. The following block of code can be thought of as a rejection-based sampler nested within the Gibbs ABC algorithm:

---

[2]For the curious reader, the file `GibbsABC.R` contains code to sample from the posterior using traditional Gibbs sampling, as well as the Gibbs ABC algorithm so that accuracy of the algorithm can be assessed. Within this code, the parameters of the conditional distribution are specified.

```
1  abc=NULL
2  abc$theta=matrix(NA,N,J)        # declare a matrix for storage
3  abc$lambda=numeric(N)           # declare a matrix for storage
4
5  for(i in 1:N){                  # loop over particles
6    for(j in 1:J){                # loop over subjects
7    d=eps+1                       # initialize d to be greater than eps
8    # continue proposal generation until condition is satisfied
9    while(d>eps) {
10     theta.1=rnorm(1,mean(Y[j,]),1)   # sample from proposal dist.
11     x=rpois(T,theta.1)                # simulate data
12     d=ifelse(theta.1>0,rho(Y[j,],x),eps+1)    # compute distance
13   }
14   abc$theta[i,j]=theta.1              # store the accepted value
15   }
16   # sample from conditional distribution of lambda
17   abc$lambda[i]=rgamma(1,J+p.lambda[1],sum(abc$theta[i,])+p.
        lambda[2])
18 }
```

Starting with Lines 1–3, we first declare some storage objects to keep track of the samples from θ and λ, and we store these samples within a list object called abc. Zooming in on Lines 7–13, we see the familiar procedure as in the rejection-based algorithm above. First, in Line 10, we sample a theta.1 from a proposal distribution. Here, we have chosen a normal distribution centered at the mean incident rate for the jth subject (i.e., centered at $\sum_{t=1}^{T} Y_{j,t}$), with a standard deviation of one. This proposal distribution is more informed than simply sampling from the prior distribution, but also does not depend on the previous states of the chains or pools of particles as in the other algorithms we have discussed. This is another type of proposal distribution that can be used effectively when summary statistics of the data map closely onto regions of the parameter space that are likely to contain the posterior distribution. Next, we simulate data X (Line 11) and compare it to the observed data Y in Line 12. In this line, we use the ifelse function to first evaluate whether theta.1 is even a possible value of the parameter (i.e., is theta.1 in the support of θ_j). As the parameters of the Poisson distribution must be positive, θ_j must be greater than zero. If this is not the case, the ifelse function will return eps+1, which is larger than d, causing theta.1 to be rejected. Also note that the comparison is made at the level of the jth subject: no other rows of Y are involved in evaluating the fitness of theta.1 (i.e., the parameter θ_j). Finally, if theta.1 is accepted, it is stored on Line 14.

The next step of the algorithm is to perform the conditional updates on the parameter λ. As we have already derived the conditional distribution of λ in Eq. (2.32), we need only draw a sample from this distribution, conditional on the current state of the chains on the ith iteration. Line 17 performs this sample by plugging in the variable abc$theta[i,] as $\theta_{1:J}$ in Eq. (2.32). On some iterations, the variable abc$theta[i,] will have a different value due to the random sampling from the conditional distribution of θ, and in so doing, the conditional distribution of λ can also be estimated.

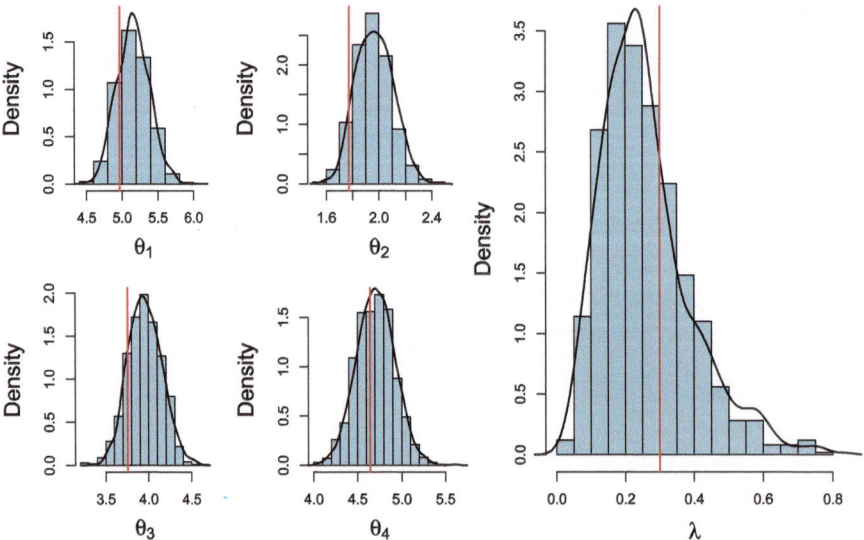

Fig. 2.8 Marginal posterior distributions for the parameters of the hierarchical Poisson example. The left panel shows the estimated posterior distributions for each θ_j parameter, whereas the right panel shows the estimated posterior for the λ parameter. Estimates obtained using the Gibbs ABC algorithm are shown as the black densities, whereas estimates obtained using a conventional Gibbs sampler (that uses the likelihood function) are shown as histograms. In each panel, the red vertical line represents the true value of the model parameters

After running the blocks of code above, we should arrive at posterior distributions that look similar to those presented in Fig. 2.8. Here, the left panel shows the estimated posterior distributions of each θ_j, whereas the right panel shows the estimated posteriors for the λ parameter. The estimates obtained using the Gibbs ABC algorithm are shown as the black densities, whereas estimates obtained using a traditional Gibbs sampler (i.e., one that uses the likelihood function) are shown as histograms. The true value of each model parameter is represented in Fig. 2.8 as the vertical line. Figure 2.8 shows that the estimated posterior obtained using Gibbs ABC closely resembles the estimates obtained using the likelihood function (i.e., the traditional Gibbs sampler). This suggests that our algorithm is working effectively to estimate the parameters for this particular hierarchical model.

2.6   Conclusions

Referring back to the introduction, successful algorithms of likelihood-free inference can be thought of as consisting of five important steps. Most of the algorithms presented here have differed only at some combination of three of these steps, although we have not reviewed every ABC algorithm that currently exists. Making different choices at these steps in the development of an algorithmic can be useful for customizing it for a particular model-fitting problem.

Fortunately, there are clear advantages and disadvantages of each choice in the development process that make one feature of an algorithm a better choice than another, depending on the circumstances. As a general guide, if sufficient statistics are known, then any of the rejection-based or kernel-based algorithms can be used to fit a model to data. Kernel-based algorithms will preserve a larger percentage of the simulated data compared to rejection-based algorithms, but also have the problem of not allowing the error term to be set to zero (as it can be in rejection-based algorithms). If the problem becomes more difficult, such as when we need to estimate many parameters, we recommend an algorithm such as ABCDE because (1) it is kernel-based and so will provide a more efficient estimate (with respect to accuracy and speed of computation time) relative to rejection-based algorithms, and (2) it uses a proposal scheme that is well suited for high-dimensional and correlated parameter spaces—a consideration important for many models in cognitive science.

If sufficient statistics are known but one would like to avoid using error terms such as ϵ or δ, then the synthetic likelihood approach is recommended. Although the synthetic likelihood method requires a great deal of more computational time, the ability to arrive at accurate posteriors and not worrying about the influence error terms may have is a desirable outcome. If no sufficient statistics are known, we recommend using the PDA algorithm as a brute force method for comparing simulated to observed data. This method, while computationally slow relative to the other methods, is a nice alternative to choosing a large set of summary statistics and hoping for the best. Finally, if one is interested in fitting a hierarchical model, we recommend the Gibbs ABC algorithm in concert with an appropriate choice for a lower-level likelihood approximation algorithm.

A Tutorial

3

3.1   Introduction

This chapter will focus on the Minerva 2 model, a global matching model of
recognition memory. Recall from Chap. 1 that the recognition memory task takes
place in at least two phases. In the first study phase, people are given a list of items
to study (e.g., words) and instructed to commit them to memory. Following the study
phase, the subject might perform some filler task, such as completing a puzzle.
Often, these filler tasks are used to either remove recency effects or to equate the
retention interval across different conditions. In the second test phase, the subject
is presented with a probe item and asked to respond either "old," meaning that the
subject believes the probe was on the previously studied list, or "new," meaning that
the subject believes the probe was not on the previously studied list. The proportion
of old responses to targets (hit rates) can be plotted as a function of the proportion
of old responses to distractors (false alarm rates), producing the receiver operating
characteristic (ROC) curve [71, 72].

After a brief description of the model, the tutorial will be broken into two
parts. In the first part, we will fit the Minerva 2 model to simulated data using
three likelihood-free methods: kernel-based Markov chain Monte Carlo, probability
density approximation (PDA), and analytic expressions derived by Sheu [37]. After
fitting the model using each technique, we will compare the three methods by
examining the estimated marginal posterior distributions for each of the model
parameters and the computational time for each method. The simulation study will
also allow us to better understand the model by permitting us to examine the joint
posteriors for each parameter generated using the PDA method.

© Springer International Publishing AG 2018   55
J.J. Palestro et al., *Likelihood-Free Methods for Cognitive Science*,
Computational Approaches to Cognition and Perception,
https://doi.org/10.1007/978-3-319-72425-6_3

In the second part of the tutorial, we will fit a hierarchical version of the Minerva 2 model to data from [4] by blending two likelihood-free techniques (PDA and Gibbs ABC). Prior to fitting the model, we will describe the hierarchical framework in detail, and after determining if the fit is adequate, we will examine the posterior distributions of specific parameters to gain a better understanding of both the effectiveness of the experimental manipulation and the predictions made by the model.

3.2   MINERVA 2 Model

Minerva 2 is a member of the class of *global matching models*, which includes memory models such as the Theory of Distributed Associative Memory model (TODAM) [73], the Search of Associative Memory model (SAM) [74], and the Matrix model [75, 76]. Global matching refers to a retrieval process in which a probe item is compared against the contents of memory, producing a single summed *familiarity* value that indexes the similarity of the probe to the contents of memory. The familiarity value is subsequently compared against a decision criterion to produce an "old" or "new" decision [77]. While all of the global matching models make mathematically identical predictions under some circumstances [78], Minerva 2 possesses some unique properties that differentiate it from the other models, such as a non-linear activation function that ensures that traces more similar to the probe have a greater contribution to the familiarity calculation.

We will describe the mathematics of Minerva 2 briefly. Readers interested in a more detailed description of the model and its predictions should consult the original publications [36, 79]. In Minerva 2, items are represented as a vector of η features that take the values of 1, 0, or -1 with equal probability. When an item is presented for study, a new trace vector is created in memory that contains features from the original item vector that are copied with probability L. The probability L is called the learning rate of the model. If a feature from the item vector is not copied into memory, the item trace has a 0 stored for that feature. After a set of items have been studied, the contents of memory are represented by a matrix M, which contains all of the trace vectors created in the study episode. Over time, correctly copied features may revert to 0 with probability δ. The probability δ is called the decay rate, and it is this decay rate that models the effects of a retention interval, with higher decay rates used for longer study-test intervals [36, 79].

Global matching operates in Minerva 2 by comparing the probe vector P against each of the trace vectors in the memory matrix M. The similarity S_i between the probe and trace vector i is calculated as

$$S_i = \sum_{j=1}^{N} \frac{P_j T_{i,j}}{\epsilon_i}, \tag{3.1}$$

where T_i is the ith trace vector in M, j is the jth feature in the comparison between the probe and trace vectors, and ϵ_i is the number of features where $P_j \neq 0$ and

$T_{i,j} \neq 0$. Values of S_i are equal to 1 if the probe vector is identical to the trace vector and 0 if the two vectors are orthogonal to each other. The extent to which a probe item activates the traces in memory is computed as the activation

$$A_i = S_i^3. \tag{3.2}$$

The cubing of the similarity values produces a non-linear relationship between similarity and activation: activation values are highest for traces that are most similar to the probe vector [79]. The activation values are subsequently summed to produce a value of "echo intensity" I (i.e., familiarity), so

$$I = \sum_{i=1}^{M} A_i. \tag{3.3}$$

The echo intensity I is compared to a decision criterion C. If I is greater than C, familiarity is high enough to produce an "old" response. If not, a "new" response is made. Because echo intensity is greatest when the similarity between the probe and the traces is highest, the values of I tend to be higher for targets: There is a higher expected similarity between the target trace vector and its own probe item than between a trace vector and an unrelated probe item. The distributions of intensity values tend to have higher variance for targets than for distractors, a difference that arises from the non-linear activation function and the probabilistic encoding of features [80]. Performance tends to be worse for longer study lists than for shorter ones because, as the number of trace vectors in M is increased, the variance of the echo intensity values increases for both targets and distractors, resulting in decreased discriminability. This list-length effect is predicted by other global matching models for similar reasons [77].

Minerva 2 and the other global matching models were challenged by a series of findings in the recognition memory literature. One of which was the *null list strength effect* [81], in which strengthening a subset of studied items by increasing study time or number of presentations does not decrease performance for the other non-strengthened items. Another was the *mirror effect* [82], whereby some manipulations produce opposite effects on the hit and false alarm rates. The global matching models were not able to capture these effects without modification, and so a newer generation of Bayesian recognition memory models were developed. These models include the Retrieving Effectively from Memory model (REM; [83]), the Subjective Likelihood in Memory model [84], and the Bind Cue Decide Model of Episodic Memory (BCDMEM; [85]), all of which can be described as the dominant theoretical models of recognition memory currently. BCDMEM and REM will become important in a later application.

Minerva 2 has been used to explain many memory-related phenomena. Arndt and Hirshman [86] found that Minerva 2 was able to successfully predict a number of relations between true and false recognition. They further demonstrated that the non-linear activation function was specifically responsible for the success of these predictions, and that other models such as TODAM or the Matrix model cannot do the same. A dual-process variant of Minerva 2 was used by Benjamin [87] to explain dissociations between item and source recognition in the aging literature. Minerva 2 was used to develop successful models of judgment and decision making, including judgments of likelihood (MINERVA DM) [88] and hypothesis generation (HyGene) [89]. It has also been applied to semantic memory phenomena, such as performance on lexical decision tasks [90] and the formation of lexical representations [91].

3.2.1   Implementing the Model

The Minerva 2 model is a simulation model, meaning that exact equations have not yet been produced for evaluating the likelihood function. As such, to fit the model to data in this chapter, we rely on likelihood-free techniques. The minimum requirement of these techniques is that we be able to simulate data from the model, which means we must first prepare computer code to implement the equations in the preceding section.

First, to offload some complexity of the model simulation code, we can write a separate function to generate the features that represent the items used at study and test.

```
init=function(N,p) sample(c(-1,0,1),N,replace=T,prob=p)
```

The `init` function simply samples the features $\{-1, 0, 1\}$ with probability determined by the variable p. The end result is a vector of randomly selected features of length N, corresponding to η. With a simple function for generating feature values, the next block of code creates a function called `minerva` that can be used to generate responses, given some parameter values and experimental variables:

```
minerva=function(L,crit,decay,n.features,alpha,p,n.study,n.
    targets,n.test){
  study.feat=init(n.features*n.study,p) # features of study list
  study=matrix(study.feat,n.features,n.study) # study list
  image.feat=rbinom(n.features*n.study,1,L) # feature of image
  image=matrix(image.feat,n.features,n.study)*study # image
  image=rbinom(n.features*n.study,1,1-decay)*image # decay
  targets=study[,1:n.targets] # use first study items (arbitrary
    )
  # calculate number of distractors, and generate them
  n.distractors=(n.test-n.targets) # how many distractors?
  dist.feat=init(n.features*n.distractors,p) # features
  distractors=matrix(dist.feat,n.features,n.distractors) # set
  test=cbind(targets,distractors) # create test set
  S=matrix(NA,n.study,n.test) # create similarity matrix
  for(i in 1:n.test){ # loop over test items
    for(j in 1:n.study){ # ...and study items
      calc.n=sum(test[,i]!=0 | image[,j]!=0) # nonzero features
      S[j,i]=sum(test[,i]*image[,j])/calc.n # similarity
```

```
18        }
19      }
20      A=S^alpha # calculate activation
21      out=apply(A,2,sum) # calculate intensity
22      as.numeric(out>crit) # compare to criterion
23    }
```

The first line of code declares the function, which requires the learning rate L (i.e., L), the criterion value crit (i.e., C), the decay rate decay (i.e., δ), the number of features n.features (i.e., η), the exponent of the similarity matrix alpha (which is commonly set to $\alpha = 3$, as in Eq. (3.2)), and the feature probability vector p (which is usually set to $p = \{1, 1, 1\}/3$, so that the features are equally likely). In addition, the minerva function requires some choices about the details of the experiment, such as the number of items presented at study (i.e., n.study), the number of items in the test list that were on the study list (i.e., n.targets), and the total number of items in the test set (i.e., n.test). Lines 2 and 3 create the features of the study list, and then arrange those features into a matrix where the columns correspond to the items, and the rows correspond to the features. Lines 4–6 detail the construction of the episodic memory matrix. First, Line 4 generates a vector of Boolean variables declaring whether or not the features of the study list should be encoded. The vector in Line 4 is rearranged into a matrix and multiplied by the study list, creating a new matrix that represents the episodic image: some features of the study list will appear within the image matrix at the same location as in the study list matrix, whereas some values within the image matrix will be zero, indicating no features were encoded. Finally, the quality of the image matrix is further deprecated by multiplying the matrix by another Boolean matrix representing the feature decay process. During this multiplication, some features that were correctly encoded in Line 5 will be reset to zero, eliminating those features from contributing to the recognition decision at test. Line 7 declares that the first n.targets items will be selected from the study list to serve as the targets in the test list. Because we are simulating a model and not using a human subject, the choice of selecting target items from the study list is completely arbitrary. Lines 9–11 create the set of distractors to be used in the test set, and Line 12 creates the final test set by combining targets and distractors. Again, because we are simulating the model, the arrangement of targets and distractors is inconsequential to the pattern of responses we will simulate.

The next step is to calculate the similarity matrix S, which is performed in Lines 13–19. To do this, we follow Eq. (3.1) by looping through the set of test items and the set of items in the episodic image. The first step is to calculate ϵ_i in Line 16 to determine how many features are nonzero in either the current test item or the current episodic image item (i.e., determined by i or j in the double for loop). Finally, Line 17 performed the summation in Eq. (3.1) through matrix multiplication. Line 20 calculates the activation values by cubing the similarity matrix shown in Eq. (3.2), and Line 21 calculates intensity according to Eq. (3.3). Finally, to make a response, the model compares the intensity of each test item to the criterion variable crit: if the intensity for Item i is larger than the criterion, an "old" response is given, whereas if the intensity value is smaller than the criterion, a "new" response is given.[1]

The Minerva 2 model is relatively simple to set up and simulate data from, and as a consequence, it serves as an interesting running example on which we can apply likelihood-free techniques to illustrate the utility of these methods. Despite Minerva 2's simplicity, to our knowledge researchers have not yet taken full advantage of Bayesian hierarchical methods in fitting the model because it is simulation-based. In the following section, we will describe how likelihood-free techniques can be used to fit this model to data. We first fit the model to simulated data to demonstrate the methods' ability to recover the model parameters. Then, we use these techniques to fit the model to recognition memory data from a real-world experiment.

3.3   Simulation Study: Recovering the Posterior Distribution

Although the Minerva 2 model can be fit to a variety of data, for our simulation study, we focus on the recognition memory task. Recognition memory data are perfect candidates for illustrating the likelihood-free approach for two reasons. First, the number of measurements (i.e., hit and false alarm rates) from each subject is generally small, and so simulating the model to match the observed data is not very computationally costly. Second, the hit and false alarm rates are discrete: measures are incremented in steps of $1/n$, where n is either the number of targets (for the hit rate) or distractors (for the false alarm rate). This means that when using methods such as PDA, the error introduced in the estimation of the posterior distributions will be minimized, as a kernel density function is not needed to approximate the shape of the probability density function [38].

3.3.1   Generating the Data

For this simulated experiment, we assumed that the test list consisted of 40 items total, 20 of which were targets (i.e., words on the previously studied list) and 20 of which were distractors (i.e., words not on the previously studied list). The study list consisted of 20 items, all of which were presented during the test phase. Each subject completed four conditions of the recognition memory task. Hence, the simulated data consist of four hit rates and four false alarm rates for each subject. The larger data set provides an opportunity for the posterior distribution to be different from the prior distribution, thus creating a greater constraint on the model. This increased stringency allows us to appreciate the quality of the likelihood approximation used by the three methods below.

---

[1]The responses are arbitrarily coded as either a one for an "old" response, or a zero for a "new" response.

To simulate data from the model, we set the learning rate $L = 0.5$, the criterion $C = 0.10$, the number of features $\eta = 30$, and the decay rate $\delta = 0$. We set the α parameter to 3 and the probability of the features taking on the values $\{-1, 0, 1\}$ to be $\{1/3, 1/3, 1/3\}$, respectively. When simulating the model for four subjects, the following hit and false alarm rates were obtained:

$$\text{Hit Rates} : \{0.40, 0.75, 0.75, 0.60\}$$

$$\text{False Alarm Rates} :   \{0.05, 0, 0, 0\}$$

This set of data will be used in the posterior recovery test below, and they are important as changes in the data above may result in changes to the posterior estimates obtained below.

For our posterior recovery test, we only estimated L, C, and η, because the analytic expressions derived in Sheu [37] did not consider the effects of the decay parameter δ. Hence, because analytic expressions for this expanded version of the model are unavailable, we assumed that δ was known to facilitate a comparison across the three methods. All parameters were equal across all four conditions of the experiment.

3.3.2   Fitting the Data

To illustrate the likelihood-free approach, we fit the model to the simulated data in three different ways. The first approach is the kernel-based ABC algorithm, which relies on summary statistics to approximate the likelihood. The second approach is the PDA method [38], which constructs an approximation to the missing likelihood via pure simulation. The third approach relies on analytic expressions derived by Sheu [37], which rely on asymptotic assumptions about the distribution of hit and false alarm rates conditional on a set of model parameters. As we will discuss in detail below, by "analytic" we mean that the approximation of the likelihood has a functional form, but we do not necessarily mean that the expressions are perfectly accurate. In order to obtain analytic expressions Sheu made some simplifying assumptions about the asymptotic properties of the distribution of echo intensities. While these assumptions are reasonable for infinitely long lists, their validity when applied to data with finite limitations has not yet been tested.

Each of the three methods is unique in the way they approximate the posterior distribution. However, when sampling from the posterior distribution, another set of algorithms are required to perturb the proposals throughout the parameter space so that an accurate posterior estimate can be achieved. To maintain consistency across the three methods, we applied an MCMC algorithm (see Chap. 2) with identical settings to obtain samples from the posterior distribution using each approximation method. Note that while the MCMC algorithm is identical, because each approximation method is different, we cannot necessarily expect that each procedure will result in identical posterior estimates; in fact, it is this comparison of posteriors that we will use to evaluate the quality of each approximation method.

While we cannot reproduce the code for the entire MCMC algorithm here, we encourage the reader to consult the online materials for versions of each method implemented in R. Again, all methods use an identical MCMC sampler to perturb proposals within the parameter space, yet have different methods for evaluating the quality of the proposal. At their core, all methods invoke a function that specifies the log likelihood of the data (i.e., the variable data), given a proposal parameter value (i.e., the variable x). The general form of the log likelihood function looks like the following block of code:

```
log.dens.like=function(x,data){
  L=x[1]; crit=x[2]; feat=x[3];  # redeclare parameters
  feat=round(feat)                # round the number of features
  if(L<=1 & L>=0 & feat>=2){     # test parameter boundaries
    ### insert specific approximation method here
    ### producing a variable called 'out'
    if(is.na(out))out=-Inf         # test for plausibility
  } else {                        # if boundary test fails...
    out=-Inf                      # reject the proposal
  }
  out # return the final log likelihood value
}
```

Line 1 through Line 12 declare the log likelihood function in R. As you can see, the function requires two inputs—the parameter proposal x and the set of data data. For convenience, Line 2 transforms the elements of the proposal vector x into the learning rate variable L, the criterion variable crit, and the number of features variable feat. Line 3 rounds the feature variable into something discrete, so that it can be used in the minerva function to construct the episodic image (i.e., only discrete values can be used as the dimensions of a matrix). Next, Line 4 tests to see whether or not each parameter value is within a plausible range. Statistically speaking, this line is not completely necessary as these restrictions will be specified in our priors. However, algorithmically, the minerva function above will crash if these restrictions are not in place. Line 4 is connected to Lines 8–10 as a condition statement. That means that if a parameter value is outside the range of plausible values, the final log likelihood value out will be set to $-\infty$ or -Inf in Line 9. Lines 5–6 represent the location where lines of code can be inserted to implement each approximation method, which we will discuss below. Line 7 is a final test to evaluate whether or not the implementation method produced a log density value that is plausible. As we assume that NA values correspond to values of the parameters that are implausible, this line is the final "catch" to rid our posteriors of invalid samples. Hence, it is very important to ensure that the approximation method—and more specifically the model data generation code—is as robust as possible. Finally, Line 11 produces the likelihood of the data, given a set of parameters, on the log scale.

With the generic wrapper function described, we can now turn to the specific implementation details that can be slotted into Lines 5–6 above.

3.3.2.1 KABC

As we discussed in Chap. 2, kernel-based ABC (KABC) relies on a kernel to compare the simulated data to the data that were observed. We chose a Gaussian kernel with standard deviation $\delta_{ABC} = 0.03$. This particular kernel worked well, giving accurate posterior estimations while still allowing the chains in the MCMC sampling algorithm to mix properly.[2] For each proposed parameter value, we simulated the model under the proposal four times to reflect the number of observed data points. Doing this allows for straightforward comparison of the simulated data to the observed data, although other choices are possible [44].

To implement the KABC algorithm, we adapted code in R based on Fig. 2.3 that would work generically across the three approximation methods. We again encourage the reader to see the scripts associated with each method, as we cannot reproduce them here. Instead, the specifics of implementing the KABC approximation can be seen in the following block of code:

```
mach=matrix(NA,S,n.test)  # declare a matrix
for(i in 1:S){  # loop over S subjects (i.e., S=4 here)
# simulate data from the model with the following settings:
# unknown parameters: L, crit, feat
# known parameters: decay, t.alpha, p
# known experimental variables: n.study, n.targets, n.test
mach[i,]=minerva(L,crit,decay=0,feat,t.alpha=3,p=c(1,1,1)*1/3,
  n.study,n.targets,n.test)
}
# calculate summary statistics for the simulated data
mach.hr=apply(mach[,1:n.targets],1,sum)/sigs            # get HR
mach.fa=apply(mach[,(n.targets+1):n.test],1,sum)/noise  # get
    FAR
# evaluate how close the simulated and observed data are
out.hr=sum(log(dnorm(mean(data$hr-mach.hr),0,.03)))  # compare HR
out.fa=sum(log(dnorm(mean(data$fa-mach.fa),0,.03)))  # compare
    FAR
out=out.hr+out.fa # calculate final 'out' variable
```

First, Line 1 declares a storage object. Line 2–9 perform a simulation using our `minerva` function from above by looping over each subject, generating data using a set of unknowns (i.e., the three parameters we are estimating) and knowns (i.e., the fixed parameters and the experimental setup), and storing the simulated responses with the matrix `mach`. Once the data have been simulated, the next step is to calculate some summary statistics. For our purposes, the set of summary statistics we chose were the hit and false alarm rates, which are calculated in Lines 11 and 12, respectively. Next, we evaluate how closely the summary statistics for the simulated data are to the observed data by using a Euclidean distance and a Gaussian kernel (as in Chap. 2) with mean 0 and standard deviation 0.03 in this case. As we are computing the log likelihood value, we can simply sum up the log-transformed

---

[2]In our simulations, we tested a few different values of δ_{ABC} until we arrived at the smallest value that still produced good mixing behavior across the chains.

likelihoods, resulting in the variables out.hr and out.fa. The final step is to obtain the log likelihood of all the data, which is obtained by summing up out.hr and out.fa in Line 16. This last line produces the variable out that can be used in the generic log.dens.like function above.

3.3.2.2 PDA

Following the details described in the previous chapter, we constructed an approximation of the joint probability density functions for hit and false alarm rates by simulating the model 1000 times for each parameter proposal. From this bivariate distribution we calculated the probability of observing the data under this parameter proposal: we evaluated the density of the constructed distribution at the location of each observed hit and false alarm rate. To implement this method, we can use the following block of code:

```
1  mach=matrix(NA,rep,n.test)   # declare a matrix
2  for(i in 1:rep){   # simulate 'rep' times (i.e., rep=1000 here)
3  # simulate data from the model with the following settings:
4  # unknown parameters: L, crit, feat
5  # known parameters: decay, t.alpha, p
6  # known experimental variables: n.study, n.targets, n.test
7  mach[i,]=minerva(L,crit,decay=0,feat,t.alpha=3,p=c(1,1,1)*1/3,
8    n.study,n.targets,n.test)
9  }
10 # calculate the hit and false alarm rates for each simulation
11 mach.hr=apply(mach[,1:n.targets],1,sum)/sigs
12 mach.fa=apply(mach[,(n.targets+1):n.test],1,sum)/noise
13 pdf=numeric(S)   # declare a storage object
14 for(j in 1:S){   # loop over subjects
15 # determine the joint probability of obtaining each observed
16 # data point, given the distribution of simulated data
17 pdf[j]=mean(mach.hr==data$hr[\,j] & mach.fa==data$fa[j])
18 if(is.na(pdf[j])==T)pdf[j]=0 # test to ensure no NA values
19 }
20 out=sum(log(unlist(pdf))) # sum up the log likelihood values
```

The PDA code is similar to the KABC code but there are some important differences. First in Line 1, the matrix mach is constructed to have rep rows and n.test columns. The difference here is that rep will be large relative to S from the above KABC code. Because S is just the number of subjects, the KABC code will generate data of the same size as the data that were observed. By contrast, the PDA code will construct a full distribution over the space of possible hit and false alarm rates. Lines 2–9 simulate the Minerva 2 model rep times and place the elements within the matrix mach. Lines 11 and 12 next complete the hit and false alarm rates for the rep simulations. The next step is to construct the simulated PDF. First, a pdf variable is constructed to contain all rep densities. Line 17 computes the joint probability that the observed hit and false alarm rates (i.e., data$hr and data$fa, respectively) match the hit and false alarm rates from the simulation above. To do this, we calculate the number of times the two elements match at the same time and then divide by the total number of observations. Another convenient way to do this is to simply take the mean of the boolean vector that performs the match comparison, as in Line 17. While this will compute the joint probability for a single data point, we must repeat this process for all of the observed data, and so the loop in Lines 14–19 is designed to carry this operation out. The final step is to check for NA values (i.e., Line 18) and construct the variable out as we did above.

3.3.2.3 Analytic Expressions

Sheu [37] derived analytic expressions for the Minerva 2 model. These expressions are based on asymptotic assumptions, and so they describe, in the limit, the mean and variance of the lure and target activation distributions. The lure and target activation distributions are assumed to be normally distributed. In our description of Minerva 2 above, we assumed that the probabilities of a feature taking on one of the values in the set $\{-1, 0, 1\}$ were equal. This is a simplifying assumption that is regularly used in practice. More generally, we can denote the probability of a feature taking on these three values as r, q, and p, respectively, where $r + q + p = 1$. In a typical application, the probability that the features take on a value other than zero is equally likely, such that $r = p$, and so $q = 1 - 2p$. Sheu [37] maintained this assumption for simplicity. Using this constraint and our notation from above, when a target is presented, the mean and variance of the activation A^+ for targets are

$$E[A^+] \approx L^3 + 3L^2(1 - L)\mu_{\eta,p}, \text{ and}$$
$$\text{Var}[A^+] \approx 9L^5(1 - L)\mu_{\eta,p},$$

where

$$\mu_{\eta,p} = \sum_{k=1}^{\eta} \left(\frac{1}{k}\right) \binom{\eta}{\eta - k} (1 - 2p)^{\eta-k}(2p)^k.$$

Although the distribution of activation for targets is easy to describe, the distribution of activation for distractors A^- is more difficult. The mean and variances of A^- are

$$E[A^-] = 0, \text{ and}$$
$$\text{Var}[A^-] = \sum_{S^-=s} s^3 p_{S^-}(s),$$

where $p_{S^-}(s)$ is the probability mass function of the distribution of similarity for distractors.

To sum across the similarity distribution for distractors, we must first calculate the echo intensity variances for distractors. Letting k be an index over all of the non-zero elements of a trace such that $0 \leq k \leq \eta$, and h be an index such that

$-k \leq h \leq k$, the probability function of the similarities of the ith item in the episodic memory matrix is

$$p_{S_i^-}(h/k) = \binom{\eta}{\eta - k} [(1 - 2p)(1 - 2Lp)]^{\eta - k}$$

$$\times \sum_{u - w = h} \binom{k}{uvw} (2Lp^2)^{u+w} [2p(1 + L - 4pL)]^v, \qquad (3.4)$$

where u, v, and w are non-negative integers constrained such that $0 \leq u, v, w \leq k$ and $u + v + w = k$. Equation (3.4) is difficult to calculate efficiently, so Sheu constructed a set of recursive equations that are easier to evaluate.

When the means and variances of both the lure and target activation distributions have been evaluated, one can compute the probability of obtaining a hit and false alarm by integrating both normal distributions from the criterion parameter C to infinity. Specifically, the hit rate H and the false alarm rate FA predicted by the model are

$$P(H|\theta) = 1 - \Phi\left(C|E[A^+], \sqrt{\mathrm{Var}[A^+]}\right) \text{ and}$$

$$P(\mathrm{FA}|\theta) = 1 - \Phi\left(C|E[A^-], \sqrt{\mathrm{Var}[A^-]}\right),$$

where $\Phi(x|a, b)$ is the normal cumulative density function at x with mean parameter a and standard deviation b. To connect these probabilities to the likelihood function, we invert the probability structure by way of multiplication as we saw in Chap. 1. We denote the number of hits for the ith subject as $O_{i,T}$ (i.e., T for targets) and the number of false alarms by $O_{i,D}$ (i.e., D for distractors), and we assume that the number of hits $O_{i,T}$ and false alarms $O_{i,D}$ arise from binomial distributions with the number of trials equal to the number of targets $N_{i,T}$ and distractors $N_{i,D}$, respectively. We can estimate the probability that the model makes an "old" response to both targets (i.e., the hit rate) and distractors (i.e., the false alarm rate) by multiplying the two probabilities together, such that

$$L(\theta|Y) = \prod_{i=1}^{S} \mathrm{Bin}(O_{i,T} \mid N_{i,T}, P(H \mid \theta)) \mathrm{Bin}(O_{i,D} \mid N_{i,D}, P(\mathrm{FA} \mid \theta)), \qquad (3.5)$$

where Y is the observed data $Y = \{O_{1:S,T}, O_{1:S,D}\}$, $S = 4$ is the number of observations, and $\theta = \{C, \eta, L\}$, and $\mathrm{Bin}(x|n, p)$ is the density of the binomial distribution, given by

$$\mathrm{Bin}(x|n, p) = \binom{n}{x} p^x (1 - p)^{n-x}.$$

To form the analytic approximation of the likelihood, we followed Sheu [37] and programmed a routine in R to evaluate the equations listed above. The code for implementing the model is provided in the software release with this book. To sample from the posterior distribution, we can insert the following block of code into the likelihood wrapper as we did in the KABC and PDA sections above:

```
# unknown parameters: L, crit, feat
# known parameters: decay, t.alpha, p
# known experimental variables: n.study, n.targets, n.test
# calculate the mean and variances from Sheu (1992)
temp=minerva2_analytic(L,feat,p)
# calculate the hit rate predicted by the model
temp.hr=1-pnorm(crit,temp$mean.signal,sqrt(temp$var.signal))
# calculate the false alarm rate predicted by the model
temp.fa=1-pnorm(crit,temp$mean.noise,sqrt(temp$var.noise))
# compute the final joint probability of the data
out.hr=sum(log(dbinom(data$hr*sigs,sigs,temp.hr)))    # HR
out.fa=sum(log(dbinom(data$fa*noise,noise,temp.fa)))  # FAR
# compute the final joint probability of the data
out=out.hr+out.fa
```

As before, the unknown parameters, known parameters, and experimental variables are listed in Lines 1–3. The first step is to pass the parameters L and η to the function minerva2_analytic which computes the mean and variances of the signal and noise distributions. Notice that this function does not depend on the experimental variables, as they reflect the long-term representations used in the model to make recognition decisions. This assumption may be problematic as Minerva 2 is a global matching model and depends on the properties of the study set. Also note that the decay term δ is not passed to the minerva2_analytic function as the likelihood approximation developed by Sheu [37] only considers the predictions of the model under the constraint that $\delta = 0$. Next, Lines 6–9 compute the hit and false alarm rates by assuming normal representations for both target and lure distributions (i.e., as in [37]). Here, the parameter C (i.e., the variable crit) is used to specify the boundary point of the integral of the normal distributions, which provides the hit and false alarm rates predicted by the model. Finally, Lines 10–14 compute the joint probability of the observed data by adding together the log densities from the binomial distribution.

3.3.3 Results

For each method, we used a generic MCMC algorithm to sample from the posterior distribution. We used a Gaussian kernel for each dimension in the parameter space and set the standard deviations to 0.1, 0.05, and 3 for L, C, and η, respectively. We ran the sampler with 24 chains for 2000 iterations, and used a burn-in period of 200 iterations. We then thinned the chains by discarding every other sample. Hence, each method produced 21,624 samples of the joint posterior distribution.

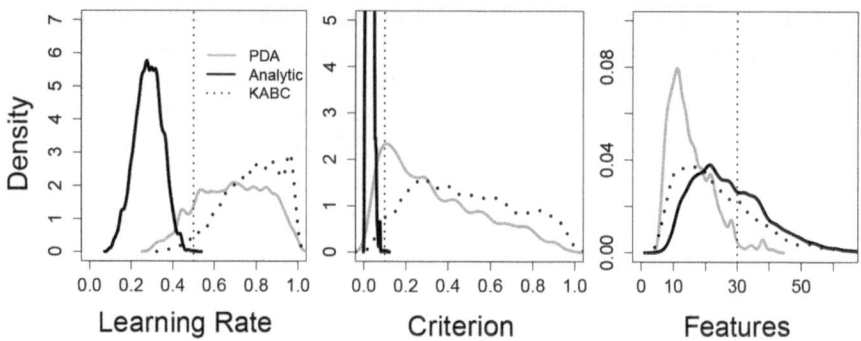

Fig. 3.1 Marginal posterior distributions for the three Minerva 2 parameters. The left, middle, and right panels show the estimated marginal posterior distributions for the learning rate L, the criterion C, and the number of features η, respectively. In each panel, three estimates are shown, one corresponding to each method used to form the approximation: PDA (gray), analytic expressions (solid black), and KABC (dotted black). In each panel, the true value of the parameter used to generate the data is shown as the vertical dashed line

Figure 3.1 shows the estimated marginal posterior distributions for each of the model parameters. The left panel shows the estimate for the learning rate L, the middle panel shows the estimate for the criterion C, and the right panel shows the estimate for the number of features η. Within each panel, the estimate obtained using each of the three methods is shown: PDA (gray), analytic expressions (solid black), and KABC (dotted black). In each panel, the true value of the parameter used to generate the data is shown as the dashed vertical line. Comparing across the three methods, Fig. 3.1 shows that the estimates obtained using PDA and KABC are more similar to one another than they are to the estimates obtained using the analytic expressions.

In particular, the estimate from the analytic expressions for the criterion parameter is highly peaked around the value 0.03 and has considerably less variance compared to the other likelihood-free methods. However, the estimated posterior is so heavily concentrated that it does not overlap with the value of the parameter used to generate the data. Similarly, the estimated posterior for the learning rate parameter using the analytic expressions also misses the true value of the parameter used to generate the data. Yet, the estimates for the learning rate parameter using KABC and PDA do overlap with the true value, and are somewhat similar to one another.

Because the true likelihood function has not been derived for Minerva 2, we cannot say for sure which of the three estimates is correct. We suspect that the inaccuracy of the posterior distributions in Fig. 3.1 reflects the effects of the normality assumption used in Sheu [37]. Although assuming that the target and lure activation distributions are normally distributed is convenient for deriving an approximation for the likelihood function, the activation distributions produced by Minerva 2 are only normal in the limit as the number of features in the trace vector increases and the number of traces in the episodic memory matrix increases. Conventional values for Minerva 2 for number of features and number of traces are around 20, and so it is possible that the approximation is less accurate under these conditions. However, because the estimates obtained from the analytic expressions miss the true value used to generate the data, we suspect that this approximation is less accurate than either the PDA or KABC methods.

The results in Fig. 3.1 also show some small differences between PDA and KABC. We suspect that the reason for the differences is due to the kernel-based approximation, which introduces some approximation error over and above the error due to Monte Carlo sampling. By contrast, the PDA method removes this approximation error, but also has the some error associated with the construction of the simulated PDF. Again, as we do not know what the true posterior distribution is, we cannot properly evaluate which estimate is more accurate (but see Chap. 4 for some validation examples).

In addition to the marginal distributions shown in Fig. 3.1, we can also examine the estimated joint posterior distributions. Occasionally, these posterior distributions reveal interesting tradeoffs in the model parameters that would otherwise be difficult to appreciate [38, 44]. Figure 3.2 shows the estimated joint posterior distribution obtained using the PDA method for each pairwise combination of the three model parameters. The left panel plots the criterion C against the learning rate L, the middle panel plots the number of features η against L, and the right panel plots η against C. Figure 3.2 reveals an interesting curvilinear pattern in the posteriors, especially in the left and right panels. Although a detailed analysis of Minerva 2 is outside the scope of this tutorial chapter, we can provide some intuition behind the tradeoff between the learning rate L and the number of features η. Recall that the memory traces are formed by a fixed number of features representing the items in the study list. After an item is presented for study, features of that item are copied into the episodic memory matrix with probability L. As L increases, the probability

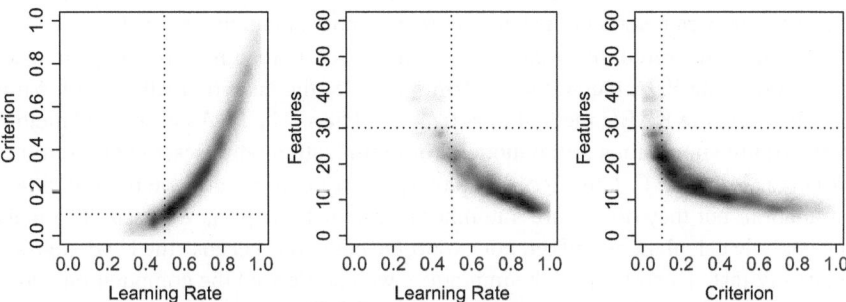

Fig. 3.2 Joint posterior distributions for the three Minerva 2 parameters. The estimated joint posterior distribution obtained using the PDA method for each pairwise combination of the three model parameters are shown: the criterion C against the learning rate L (left), the number of features η against L (middle), and η against C (right). The true values of the parameters used to generate the data are shown as the dashed lines

of copying a given feature increases, making recognition performance better at test. Hence, when the learning rate is low, Minerva 2 needs more features of the item to accurately copy enough features to maintain the same accuracy observed in the data. A similar pattern of tradeoffs exists in the Retrieving Effectively from Memory (REM) [44, 83, 92] model.

We can learn a good deal about model constraint, flexibility, and identifiability by close examination of the model parameters. For example, in typical applications, we assume a diffuse prior on the model parameters. After the posterior distribution has been estimated, we can compare the spread of the prior distribution relative to that of the posterior. If a significant discrepancy is observed, where the posterior distribution has smaller variance than the prior distribution, we can conclude that the data are constraining the model, because the data have provided evidence that reduces our uncertainty about the parameter values. Hence, the spread of the joint posteriors in Fig. 3.2 suggests that the parameters are well identified, and we learn a lot from our data. In addition, the parameters are highly correlated and trade off against each other, a finding that is often observed in computational psychological models [38, 44, 57]. Finally, the joint posteriors reveal reasonably good accuracy with respect to recovering the true parameter values (also see Fig. 3.1), which further suggests the model is well identified.

In general, if we were to assess whether or not the posterior estimates were accurate, we would need the true likelihood function so that we could compare the estimates obtained by the approximation methods. However, as we don't currently have the likelihood function for the Minerva 2 model, we must instead compare the estimated posteriors to the true value used to generate the data. Unfortunately, the comparison is not as simple as evaluating the density of the true value of the parameter within the posterior distribution. In fact, there is nothing that guarantees an estimated posterior distribution will center around a true parameter value in these types of simulation studies. All we can assess is whether or not the true value is contained within the posterior. For KABC and PDA, all of the true values are contained somewhere within the posterior. However, for the analytic expressions, the criterion parameter and the learning rate do not contain the true value.

We also measured the total computational time required to complete each simulation. The PDA method took 10 min and 2 s, the analytic method took 6 min and 26 s, and the KABC method took 6 min and 58 s. The PDA and KABC method both require simulations of the model. For optimization purposes, we programmed Minerva 2 in the C language. The analytic expressions do not require model simulation, but they do require a number of calculations [37]. We used R to make the analytic calculations and did not export the code to C because the R version was relatively fast. To perform each simulation, we parallelized the computation across 8 cores on a Mac Pro desktop computer with a 3.7 GHz processor. The computation times reveal an interesting tradeoff. First, the fastest results are obtained using the analytic expressions, and the slowest using the PDA method. However, we believe that the analytic expressions are also the most inaccurate, and the PDA method is the most accurate as no error terms corrupt the estimates. However, the differences between the PDA and KABC estimates are not large, which suggests that the KABC algorithm may be a suitable approximation for more intensive model fits.

3.4   Real-World Application: Dennis et al. [4]

Our next exercise fits a hierarchical version of Minerva 2 to data from Dennis et al. [4]. In this experiment Dennis et al. [4], had people perform a recognition memory task that manipulated list length and the presence or absence of an additional filler task between the study and test phase of the experiment. In addition, they implemented a number of controls to eliminate confounds present in traditional list length designs. The first of these confounds is an unequal retention interval across short and long list length conditions, which they controlled by keeping the time between the onset of the study list and the onset of the test list the same in both conditions. The second is the possibility of a decrease in attention which might occur over the presentation of the study list. They controlled this possibility by only testing items from the beginning of the study list in both short and long list conditions.

Dennis et al. [4] found that the list length effect depended on the presence of the filler task. Specifically, when no filler task followed the study list, recognition performance was better for short lists than for long lists. However, when a filler task was present there was no list length effect. We chose to fit the model to this particular dataset because the two independent variables, list length and retention interval, are both variables for which Minerva 2 makes predictions.

3.4.1   The Model

To fit Minerva 2 to the data from Dennis et al. [4] we must first recall the roles that each parameter in the model plays and their relationships to the experimental design.[3] We discussed already how Minerva 2, as other global matching models, accounts for effects of list length as a function of the number of trace vectors in the memory matrix. Specifically, the matching process comparing an item from the test list to the episodic image involves a comparison to every item from the study list. Next, we must consider how the effects of the filler task might influence memory performance in each task, and how the model might account for these changes across tasks. We assumed that the effects of the filler task can be explained by the decay rate parameter δ. We assume that, in the presence of a filler task, the individual traces in memory decay with probability δ, an assumption that is consistent with the original implementation of the model [79]. Recall that if a feature in a trace decays that it reverts to a zero in the episodic image. As a result, increases in δ produce more memory decay, resulting in lower discriminability.

To implement this mechanism we define a binary indicator F_j to designate the parameters for a given condition j. We let F_j represent the condition in which the

---

[3]Dennis et al. [4] also used words of different frequency (high and low) to construct their study and test lists. Because Minerva 2 lacks a mechanism for explaining word frequency effects in recognition memory, for the purposes of this demonstration we collapsed across both word frequency classes to produce a single hit and false alarm rate for each experimental condition.

additional filler activity was either absent ($F_j = 0$) or present ($F_j = 1$). We can then write the parameter vector $\theta_{i,j}$ for the ith subject in the jth condition as

$$\theta_{i,j} = \{L_i, \delta_i F_j, \eta_i, C_i^{(j)}\}. \tag{3.6}$$

Thus, in conditions with no filler task ($F_j = 0$), the decay parameter $\delta_i = 0$, but when a filler task is present ($F_j = 1$), $0 \le \delta_i \le 1$. The assumption setting the decay rate is zero when no filler task is present is an arbitrary one, as the other model parameters should scale accordingly. The important difference here is the value of the decay rate δ_i relative to zero.

For the ith subject in the jth condition, we denote the number of hits as $O_{i,j,T}$ and the number of false alarms by $O_{i,j,D}$. The number of hits $O_{i,j,T}$ and false alarms $O_{i,j,D}$ arise from binomial distributions with the number of trials equal to the number of targets $N_{i,j,T}$ and distractors $N_{i,j,D}$, respectively. We can estimate the probability that the model makes an "old" response to both targets (i.e., the hit rate) and distractors (i.e., the false alarm rate) by simply simulating the model many times and tabulating the responses under the different stimulus types. These values give us the probability of a hit $P(H \mid \theta_{i,j})$ and false alarm $P(\text{FA} \mid \theta_{i,j})$. Letting the observed data

$$Y = \{\{O_{1,1,T}, O_{1,1,D}\}, \{O_{1,2,T}, O_{1,2,D}\}, \dots, \{O_{1,J,T}, O_{1,J,D}\},$$
$$\{O_{2,1,T}, O_{2,1,D}\}, \{O_{2,2,T}, O_{2,2,D}\}, \dots, \{O_{2,J,T}, O_{2,J,D}\},$$
$$\dots,$$
$$\{O_{S,1,T}, O_{S,1,D}\}, \{O_{S,2,T}, O_{S,2,D}\}, \dots, \{O_{S,J,T}, O_{S,J,D}\}\}$$

for J conditions and S subjects, the likelihood function for Y is

$$L(\theta \mid Y) = \prod_{i=1}^{S} \prod_{j=1}^{J} \text{Bin}(O_{i,j,T} \mid N_{i,j,T}, P(H \mid \theta_{i,j})) \text{Bin}(O_{i,j,D} \mid N_{i,j,D}, P(\text{FA} \mid \theta_{i,j})), \tag{3.7}$$

where $S = 48$ is the number of subjects, and $J = 4$ is the number of conditions.

To complete the model we select priors for each of the parameters. Because the parameters L_i and δ_i represent probabilities, they are both restricted to be between zero and one. We therefore assumed that each of the individual-level parameters δ_i and L_i have truncated normal priors bounded by zero and one, or

$$L_i \sim \mathscr{T}\mathscr{N}(\omega_L, \xi_L, 0, 1), \text{ and}$$
$$\delta_i \sim \mathscr{T}\mathscr{N}(\omega_\delta, \xi_\delta, 0, 1),$$

where $\mathscr{T}\mathscr{N}(a, b, c, d)$ denotes the truncated normal distribution with mean a, standard deviation b, lower bound c, and upper bound d. Because the number of features parameter η_i can only take on integer values, we used a discretized version of the truncated normal distribution with mean parameter ω_η, standard deviation parameter ξ_η, and with a lower bound of two so that

$$\eta_i \sim \mathscr{D}\mathscr{T}\mathscr{N}(\omega_\eta, \xi_\eta, 2, \infty),$$

where $\mathscr{D}\mathscr{T}\mathscr{N}(a, b, c, d)$ denotes the discretized truncated normal distribution with mean a, standard deviation b, lower bound c, and upper bound d. Choosing a lower bound of two restricted each item to have at least two features.[4]

Unlike current Bayesian recognition memory models [83, 85], Minerva 2 and the other global matching models use different criteria for an "old" response for different experimental conditions. This is because changes in Minerva 2 parameters, such as the learning rate and decay rate, affect the mean of the target activation distribution without changing the mean of the distractor activation distribution (although the variances of these distributions are different). To illustrate why a single criterion is insufficient, consider two experimental conditions with different retention intervals, or the length of time between study and test phases of the experiment. The longer retention interval condition will have a higher decay rate, which will decrease the mean of the target activation distribution. If the decision criterion is fixed for long and short retention intervals, the lower mean target activation will result in lower hit rates, but the false alarm rate will not be affected. If the decision criterion is reduced in the long retention condition, the hit rate will be lower and the false alarm rate will be higher than in the short retention condition, which is consistent with experimental findings. Allowing for different decision criteria across conditions also allows for unbiased responding in each condition.

To ensure that the model can fit the data, we assumed a separate criterion parameter $C_i^{(j)}$ for each of the $J = 4$ conditions. This is not without precedent, as a similar approach was used by Clark and Shiffrin [93] to fit Minerva 2 to their recognition data. For each individual criterion we specified a normal prior distribution, so that

$$C_i^{(j)} \sim \mathscr{N}\left(\omega_C^{(k)}, \xi_C^{(k)}\right).$$

Because Minerva 2 has never been fit hierarchically to data, we have no information about the likely ranges of the hyperparameters. As a consequence, we used noninformative priors for the hyperparameters to reflect this uncertainty. First, for the mean parameters ω_L and ω_δ, we chose a uniform distribution that put equal density on all values in the interval (0,1), so that

$$\omega_L \sim \text{Beta}(1, 1), \text{ and}$$

$$\omega_\delta \sim \text{Beta}(1, 1).$$

---

[4]After fitting the model, we noticed that no marginal distribution for η_i went below four, so while our choice was made out of convenience, it had little effect on the posterior estimates.

To choose priors for the criteria and number of features, we examined the model predictions under a variety of different choices for $\{\eta, C\}$. We found that values of C ranging from 0 to 0.5 with the number of feature ranging from 10 to 20 produced data that one might expect from a typical recognition memory task. Thus, we settled on mildly informative priors for these parameters, given by

$$\omega_C^{(k)} \sim \mathcal{N}(0.05, 1), \text{ and}$$

$$\omega_\eta \sim \mathcal{TN}(40, 15, 2, \infty).$$

For the standard deviation hyperparameters $\xi = \left\{ \xi_L, \xi_\delta, \xi_C^{(k)}, \xi_\eta \right\}$, we used a common, mildly informative priors, so that

$$\xi \sim \Gamma(1, 1), \quad \cdot$$

We chose these priors because we expected only a moderate degree of variability in the individual-level parameters, and the $\Gamma(1, 1)$ distribution covered a sufficiently large range.

Figure 3.3 shows a graphical diagram for this hierarchical model. These types of diagrams are often very useful for illustrating how the parameters in the model (white nodes) are connected via arrows to the observed data (gray nodes) [6, 12, 17]. When the variables are discrete they are shown as square nodes, whereas when the variables are continuous they are shown as circular nodes. A double bordered variable indicates that the quantity is deterministic, not stochastic, and computed from other variables. For example, the node corresponding to $\theta_{i,j}$ is double bordered because it is always determined by evaluating Eq. (3.6). Finally, "plates" show how vector-valued variables are interconnected. For example, the node for the parameter ω_L is not within the plates, which indicates that this parameter is fixed across both subjects and conditions, whereas there are separate L_i nodes for every subject, and separate $C_i^{(j)}$ nodes for each subject and condition.

3.4.2 Results

To fit the hierarchical model we used the PDA method [56] embedded within the Gibbs ABC algorithm [70]. We generated proposals using DE-MCMC [57]. In the implementation of the algorithm we ran 24 chains in parallel, used a burn-in period of 3000 iterations, and then ran the sampler for 3000 more iterations. Although each chain was individually assessed for convergence, estimates were formed by collapsing across all 24 chains and all 3000 samples, resulting in 72,000 samples of the joint posterior distribution.

To get a sense of whether the model fits the data well, we examined the posterior predictive distribution (PPD) and compared it to the observed data. The PPD is the marginal distribution of new, unobserved data given the data already collected. It gives a prediction about how new data will be distributed if we were to collect more. In a hierarchical model, we can generate the PPD on a subject level or a group level.

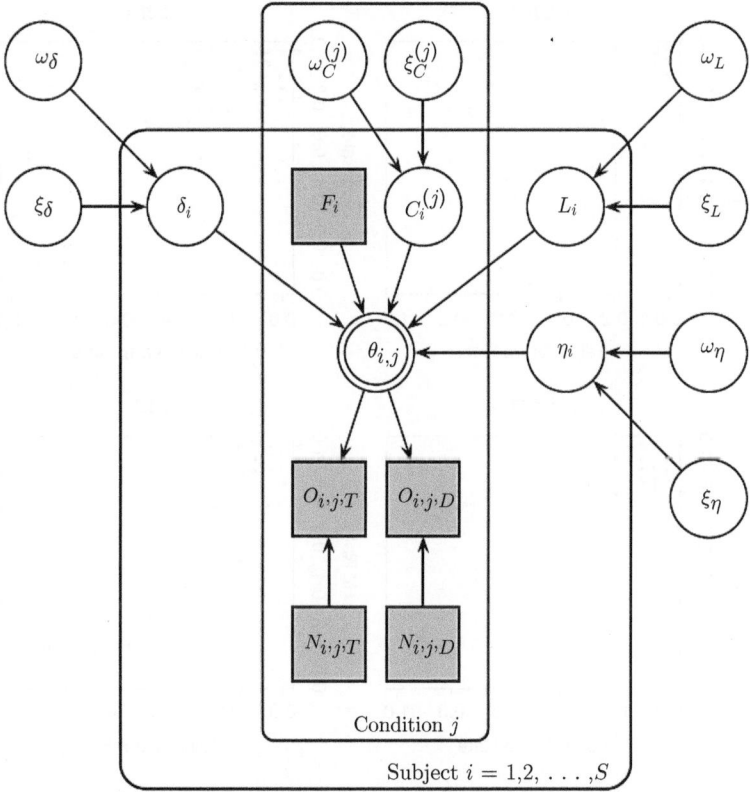

Fig. 3.3 Graphical model for the hierarchical version of the Minvera 2 model fit to the data of [4]. Parameters in the model are represented as white nodes, observed data variables are represented as gray nodes, and deterministic variables are represented as double-bordered nodes. Discrete variables are represented as square nodes, whereas continuous variables are circular. Plates illustrate a replication of a structure within the model, such as parameters across subjects or across conditions

Figure 3.4 shows the PPD at the group level separated across the four conditions. The first column corresponds to conditions in the experiment when no filler was present (i.e., $F0$), whereas the second column corresponds to conditions when the filler was present (i.e., $F1$). The rows correspond to the two list length conditions, where $L0$ denotes the short list condition (i.e., 20 words), and $L1$ denotes the long list condition (i.e., 80 words). In each panel, the black dots correspond to the observed data from Dennis et al. [4] and the gray densities correspond to the PPD. Figure 3.4 shows that the PPD is extremely variable, spreading across the majority of the ROC space. However, this is also true of the observed data [94]. Figure 3.4 assures us that the predictions of the model, which are derived from the fits, are at least sensible in that none of the observed data points fall in a location in the ROC space that is not predicted by the model.

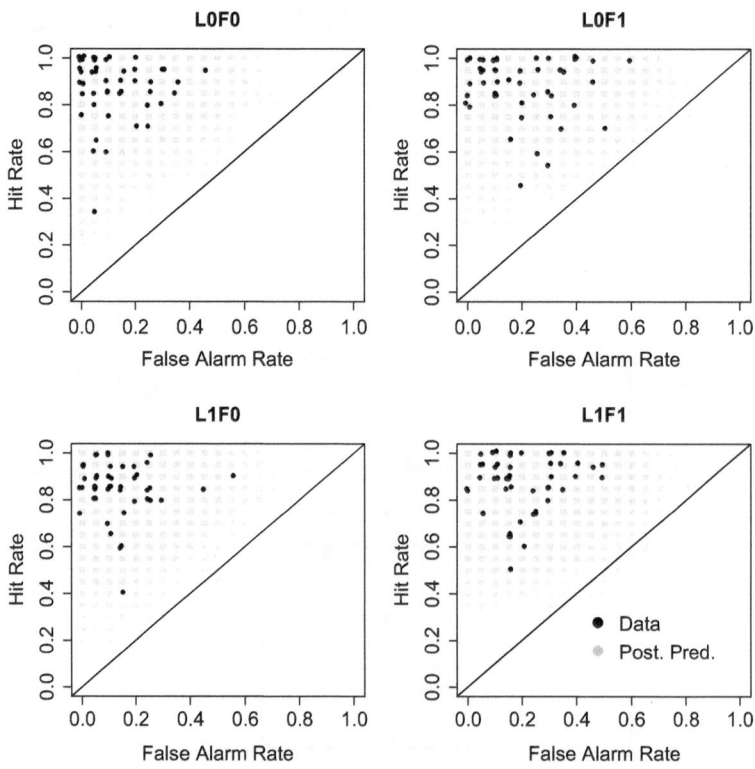

Fig. 3.4 The posterior predictive distributions (PPD) from the hierarchical Minevera 2 model. Each of the panels corresponds to a condition in the experiment from [4]: the columns correspond to the two filler conditions (i.e., filler absent condition $F0$ in the left column, and filler present condition $F1$ in the right column) whereas the rows correspond to the two list length conditions (i.e., the short list condition $L0$ in the top row, and the long list condition $L1$ in the bottom row). Observed data are represented as the black dots, whereas the gray density represents the PPD

Having assured ourselves that the model was fitting the data properly, we examined the posterior distributions. Although there are many parameters we could inspect, we focused on the group level hypermean parameters. Figure 3.5 shows the estimated posterior distributions for the four criterion parameters (top row), the learning rate parameter (bottom left), the decay parameter (bottom middle), and the number of features (bottom right). Figure 3.5 shows that the learning rate parameter for these subjects was quite high, with a mean of 0.955. The decay parameter was very low, with a mean of 0.019. Together these two parameters are important determinants of overall accuracy in the model, and these values in particular help to fit the data (see Fig. 3.4). The number of features parameter is relatively low, having a mean of 8.47. However, from our simulation study above, we now know that when the learning rate parameter is high, fewer numbers of features are required for the model to capture high accuracy data (see Fig. 3.2).

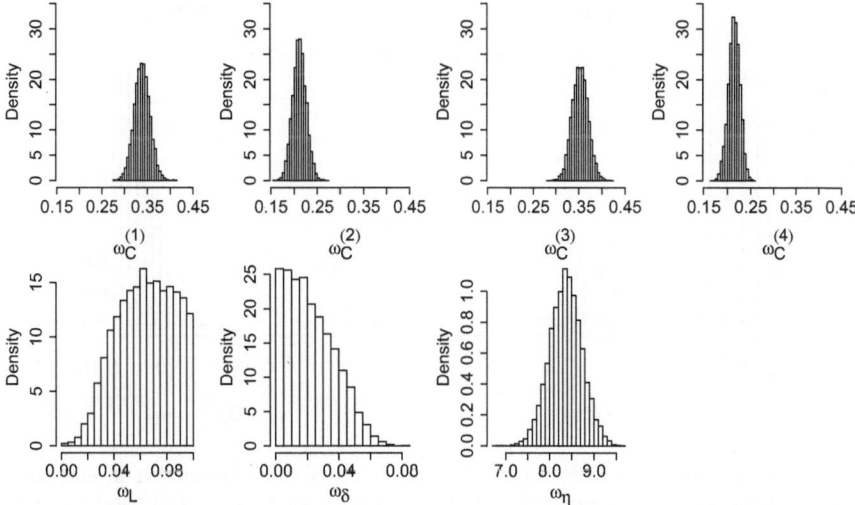

Fig. 3.5 Hypermean parameters for each of the parameters in the hierarchical Minerva 2 model. The top row plots each of the four criterion parameters $\omega_C^{(j)}$, whereas the bottom row plots the posteriors for the learning rate parameter ω_L (left), decay parameter ω_δ (middle), and the number of features parameter ω_η (right)

The two most important parameters to explain the experimental effects are the decay rate δ and the criterion parameter C. Figure 3.6 shows the estimated posterior distributions of the group-level decay rate (left panel) and the difference in the average criterion parameters (right panel). The average criterion difference was computed by collapsing over the criteria for the different list length conditions. Collapsing was justified because Fig. 3.5 shows the posterior means and variances of the decay rates are similar in the first and third conditions (short and long list lengths for no filler task) and in the second and fourth conditions (short and long list lengths for the filler task). If there was no difference in the activations across filler conditions, the estimated posterior of the average criterion difference should be centered at zero. However, Fig. 3.6 shows that the estimated posterior distribution of the difference has a mean of 0.13 and a 99% probability interval that does not contain zero. This suggests that, for the filler conditions, although the decay rate was low, it affected the activations to the extent that the criterion parameter had to shift to maintain the correct pattern of hit and false alarm rates.

We can use the estimated model parameters to gain some insight into the activation distributions used by the subjects. To do this, we generated predictions for the activations in the model for targets and distractors, conditional on the estimates of the group-level parameters (see Fig. 3.5). Figure 3.7 shows the distributions for targets (red) and distractors (gray) when the filler task is present (right panel) and absent (left panel). The criterion parameter is shown as the solid vertical line. Examining the distributions across the conditions provides better insight into why

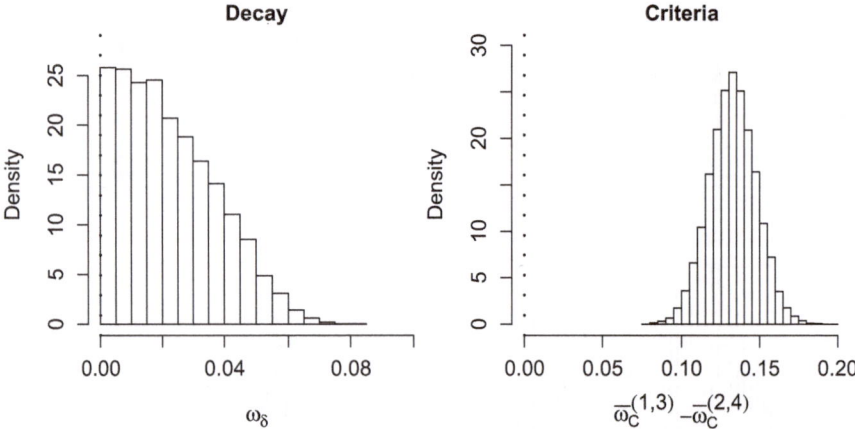

Fig. 3.6 Estimated posterior distributions of the parameters involved in capturing the filler effect. The left panel shows the group-level decay rate parameter, whereas the right panel shows the difference in the average criterion parameters in the two filler conditions. In both panels, the vertical dotted reference lines indicate the value of the parameter that would produce no filler effect

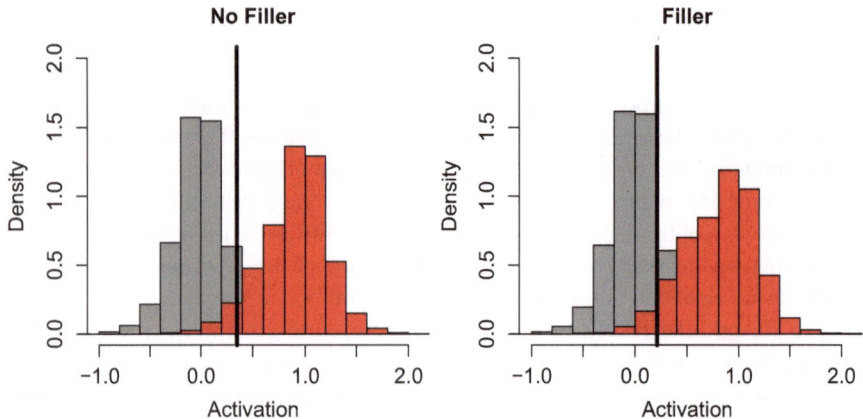

Fig. 3.7 Target and lure activation distributions used by the model across filler conditions. Target activation distributions are illustrated by the red histograms, whereas lure activation distributions are illustrated by the gray histograms. The black vertical line represents the criterion parameter used in each condition

the criterion is lower in the filler condition. Specifically, when the filler task is present, the target representation decays away, and the target activation distribution moves closer to the lure distribution. As a consequence, the criterion parameter needs to adjust to maintain a hit rate that is consistent with the observed data. Hence, together Figs. 3.6 and 3.7 illustrate an important interaction that occurs between the criterion parameters and the decay rate parameter.

3.5 Summary

In this chapter, we illustrated how likelihood-free techniques can be applied to the Minerva 2 model of recognition memory. We began by estimating the joint posterior distribution of the model parameters by generating synthetic data from the model and fitting the model to these data. We compared the estimates of the joint posterior distribution obtained using three different techniques: probability density approximation (PDA) [38], analytic expressions [37], and kernel-based ABC (see Chap. 2). We showed that while the estimates obtained using the two likelihood-free approximations converged to similar values, the analytically convenient approximations diverged from the other methods. Finally, we applied a hierarchical version of Minerva 2 to data from [4] and examined the estimated posterior distributions of the model's hyperparameters. This exercise shows how useful the likelihood-free techniques can be for the Minerva 2 model, a model that has never been incorporated into a hierarchy or fit to data using Bayesian techniques.

Validations

4

4.1 Introduction

In Chap. 3 we used the Minerva 2 model as a basis for a tutorial on the use of likelihood-free methods. We fit Minerva 2 to data, and compared the estimated posteriors using likelihood-free algorithms to those obtained using analytic expressions derived in Sheu [37]. Although these analyses presented interesting case studies about how we can use likelihood-free algorithms to estimate the model's parameters, as the true likelihood of the Minerva 2 model has not been derived for the general case, we have not yet demonstrated that the algorithms discussed in this book can provide estimates from the correct posterior distribution.

In this chapter, we show a few examples where likelihood-free algorithms have been used to recover posterior distributions correctly. First, we show how the parameters of the Bind Cue Decide Model of Episodic Memory (BCDMEM) [85] can be accurately recovered using the ABCDE [56] algorithm. This simulation study was carried out and reported in Turner et al. [92], and so we refer the reader to this work for more details on the model and simulation study reported below. Second, we show how parameters of a hierarchical signal detection theory [71] model can be recovered using the Gibbs ABC algorithm [70] and a kernel-based approach [55]. This simulation study was carried out in Turner and Van Zandt [70], and so we refer the reader to this work for details on the analyses we highlight below. Finally, we show how the parameters of the Linear Ballistic Accumulator [95] model can be accurately estimated using the PDA method [38], but not using the synthetic likelihood approach [43]. This simulation study was carried out in Turner and Sederberg [38], where the reader can find more details about the analysis.

© Springer International Publishing AG 2018
J.J. Palestro et al., *Likelihood-Free Methods for Cognitive Science*,
Computational Approaches to Cognition and Perception,
https://doi.org/10.1007/978-3-319-72425-6_4

4.2   Validation 1: The Bind Cue Decide Model of Episodic Memory

The BCDMEM postulates that when a probe item is presented for recognition, the contexts in which that item was previously experienced are retrieved and matched against a representation of the context of interest. BCDMEM consists of two layers of nodes. The input layer represents items in a local code: each node corresponds to one item. The output layer represents contexts in a distributed code: a pattern of activation over a set of nodes. When an item is studied, a random context pattern of length v for that study episode is constructed by turning on nodes in the output layer with probability s (the context sparsity parameter). The node in the input layer representing the studied item is connected to the nodes in the output layer through associative weights. These connections are established during study by connecting the active nodes on the input and the context layers with probability r (the learning rate).

During the test phase, the presentation of a probe results in the activation of the corresponding node at the input layer. This node then activates a distributed pattern of activity at the output layer that includes both the pre-experimental contexts in which the item has been encountered, which activate nodes with probability p (the context noise parameter), and the context created during study if the item appeared of the study list. This pattern is called the retrieved context vector.

The presentation of a probe also causes the reconstruction of a representation of the study list context called the reinstated context vector. The reconstruction process is unlikely to be completely accurate: nodes that were active during the study phase may become inactive with probability d (the contextual reinstatement parameter).

When a person is asked whether he has seen an item before, he bases his "old" decision on a comparison between the activation patterns of the reinstated and retrieved context vectors. As in Dennis and Humphreys [85], the ith node in the reinstated context vector is denoted by c_i and the jth node in the retrieved context vector is denoted by m_j. Both c_i and m_j are binary, indicating that the nodes i and j are either inactive or active, so $c_i = 0$ or 1 and $m_j = 0$ or 1, respectively.

To evaluate the match between the reinstated and retrieved context vectors, we let $n_{i,j}$ denote the number of nodes in the reinstated context vector that are in state i (0 or 1) at the same time that the nodes in the retrieved context vector are in state j (0 or 1). For example, $n_{1,1}$ denotes the number of nodes that are simultaneously active in both the reinstated and retrieved context vectors. Similarly, $n_{0,1}$ denotes the number of nodes that are inactive in the reinstated context vector but active in the retrieved context vector. We can then compute the probability that a probe item is a target and contrast that with the probability that a probe item is a distractor by computing a likelihood ratio given by

$$\mathscr{L}(\mathbf{n} \mid \theta) = \left[\frac{1 - s + ds(1 - r)}{1 - s + ds} \right]^{n_{0,0}} \left[\frac{r + p - rp}{p} \right]^{n_{1,1}} (1 - r)^{n_{1,0}}$$

$$\times \left[\frac{p(1-s) + ds(r + p - rp)}{p(1-s) + dsp} \right]^{n_{0,1}}, \tag{4.1}$$

where $\theta = \{d, p, r, s, v\}$ is the set of parameters for BCDMEM and $\mathbf{n}$ represents the vector of frequencies of node pattern matches and mismatches, so $\mathbf{n} = \{n_{0,0}, n_{0,1}, n_{1,0}, n_{1,1}\}$.

When we use likelihood-free approaches to estimate the posteriors of the model parameters, we are always concerned about the accuracy of those posteriors. One way to evaluate accuracy is to try to recover the parameters that were used to simulate a data set by fitting the model and comparing the estimated posteriors to the known values of the parameters. In addition, because the likelihood function has been derived for BCDMEM [96], we can evaluate whether the estimates of the posterior distributions obtained using likelihood-free methods are similar to the estimates obtained using standard likelihood-based techniques. If the two estimates are similar, then we can declare that the likelihood-free method that was used provides an accurate posterior estimates for this model.

Equation 10 in [96] provides the explicit likelihood function for BCDMEM as a system of equations. We will refer to this likelihood as the "exact" equations. Unfortunately, the exact equations can be difficult to evaluate precisely for all values of θ. For this reason, Myung et al. [96] also derived asymptotic expressions (their Equations 15 and 16) that approximate the exact solution. We will refer to this second set of equations as the "asymptotic" equations. The exact and asymptotic expressions for the hit and false alarm rates allow us to use standard MCMC methods to estimate the posterior distribution for the parameter set θ so long as v is fixed to some positive integer (v must be fixed or the other parameters are not identifiable) [96].

4.2.1 Generating the Data

To perform our simulation study, we first generated data from the BCDMEM for a single person in a recognition memory experiment with four conditions. In each of the four conditions, the simulated person was given a 10-item list during the study phase. At test, the person responded "old" or "new" to presented probes according to whether it was more likely that the probe was a target or distractor. The test lists consisted of 10 targets and 10 distractors.

To both generate and fit the data, we fixed the vector length v at 200 and the context sparsity parameter s at 0.02. We then generated 20 "old"/"new" responses for each condition using $d = 0.3$, $p = 0.5$, and $r = 0.75$. With v and s fixed, our goal was to estimate the joint posterior distribution for the parameters d, p, and r.

4.2.2   Recovering the Posterior

To assess the accuracy of our likelihood-free approach, we fit the model to the simulated data in three ways. First, we used the ABCDE [56] algorithm to approximate the likelihood function. As discussed in Chap. 2, the ABCDE algorithm is a kernel-based approach, and so we specified a normal kernel function where the summary statistics of the data were the hit and false alarm rates. The spread of the kernel function was fixed to $\delta_{ABC} = 0.1$ through the estimation process. Second, we used the standard Bayesian approach by using the exact expressions of the likelihood function from Myung et al. [96]. Third, we used the asymptotic expressions of the likelihood function from Myung et al. [96]. As we did in the simulation study from Chap. 3, the details of the sampling algorithm were fixed across the three likelihood approaches, and only the code corresponding to the evaluation (or approximation) of the likelihood function was changed across the three methods. We used DE-MCMC [57,60] as the sampling algorithm and obtained 10,000 samples in total. After discarding a burn-in period of 1000 samples, we were left with 9000 samples collapsed across 12 chains. We used standard techniques to assess convergence of the chains (using the coda package in R) [97,98].

4.2.3   Results

Figure 4.1 shows the estimated marginal posterior distributions for d, p, and r using ABCDE (the gray lines), exact (solid black lines), and the asymptotic (dashed black lines) expressions. Each panel includes the distributions obtained for a single parameter, and the dashed vertical lines indicate the true value of that parameter that generated the data.

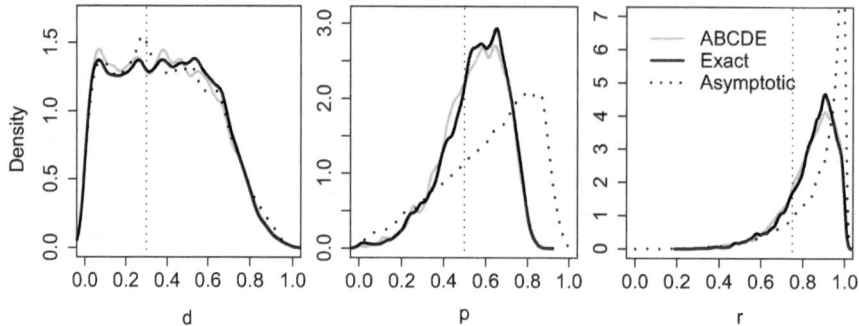

Fig. 4.1 The approximate marginal posterior distributions for the parameters d (left panel), p (right panel), and r (right panel) in BCDMEM using ABCDE (gray lines), the exact likelihood expressions (solid black lines), and the asymptotic likelihood equations (dashed black lines). The dashed vertical lines are located at the values of the parameters used to generate the data

There are two main results of our simulation study. First, the estimated posteriors we obtained using ABC are very similar to those obtained using the exact expressions for the likelihood. This suggests that the combination of $\rho(X, Y)$, $\mathcal{K}$, and δ we selected produced accurate ABC posterior estimates. Second, the posterior estimates we obtained using the asymptotic expressions are different from those we obtained with the true likelihood, especially the posterior estimates for the parameters p and r. This inaccuracy suggests that the asymptotic expressions will not be very useful for Bayesian analyses of BCDMEM.

In addition to the differences in posterior estimates that we obtained with each method, the computation times required to obtain these estimates also varied considerably with different methods. The method of estimation using exact likelihoods required 2 h and 20 min of computation. The method using the asymptotic expressions took only 36 s. The ABCDE approach took 2 min and 33 s. While the asymptotic expressions did provide the fastest results, they were considerably less accurate compared to the ABCDE approach. Perhaps more interesting is that the ABCDE approach was 55 times faster than when using the exact expressions, which we take as a testament to the usefulness of these kernel-based approaches for fitting simulation models such as BCDMEM.

4.2.4 Summary

In this section, we illustrated the utility of the ABC approach by fitting the model BCDMEM to simulated data. The derivations in Myung et al. [96] provided expressions for the model's likelihood, which allowed us to compare estimates of the posterior distribution obtained with standard Bayesian techniques to the estimates obtained with likelihood-free techniques. We showed that the estimates obtained using ABCDE were very close to the estimates obtained using the exact expressions, but the estimates obtained using the asymptotic expressions did not closely match either the ABCDE or exact expression estimates.

4.3 Validation 2: Signal Detection Theory

Signal detection theory (SDT) is one of the most widely applied theories in all of cognitive psychology for explaining performance in two-choice task. In these tasks, someone is presented with a series of stimuli and asked to classify each one as either signal (a "yes" response) or noise (a "no" response). What constitutes noise and signal can be flexible. For example, a person may be asked to indicate whether they have observed a flashing light by responding "yes" if they've detected it or "no" if they have not. The variability in the sensory effect of the stimulus, due either to noise in the person's perceptual system or to variations in the intensity of the signal itself, is represented by two random variables: the first is the sensory effect of noise when no light is presented, and the second is the sensory effect of signal when the light is presented. Typically, a presentation of a signal (a flashing light) will result in larger sensory effects than the presentation of noise alone.

The psychological representations of the effects of signals and noise are frequently modeled with two random variables. These variables are assumed to be normally distributed with equal variance, although neither of these assumptions is necessary. The equal-variance, normal version of the SDT model has only two parameters. The first parameter d represents the discriminability of signals and is the standardized distance between the means of the signal and noise distributions. Higher values of d result in greater separation and less overlap between the two distributions, meaning that signals are easier to discriminate from noise. The second parameter is a criterion c, which is along the axis of sensory effect. A person makes a decision by comparing the perceived sensation to c. If the perceived magnitude of the effect is above this criterion, the person responds "yes." If not, the person responds "no" [99].

When the two representations (signal and noise) have equal variance and the payoffs and penalties for correct and incorrect responses are the same, an "optimal" observer should place his or her criterion c at $d/2$. This is the point where the two distributions cross, or equivalently the point at which the likelihoods that the stimulus is either signal or noise are equal. We can then write the "non-optimal" observers criterion c as $d/2 + b$, where b represents bias, or the extent to which the person prefers to respond "yes" or "no." Negative bias shifts the criterion toward the noise distribution, increasing the proportion of "yes" responses, while positive bias shifts the criterion toward the signal distribution, increasing the proportion of "no" responses.

SDT has been influential because it separates effects of response bias from changes in signal intensity. The parameter d, the distance between the means of the two representations, increases with increasing stimulus intensity. The parameter b, the person's bias, is an individual-level parameter that can be influenced by instructions to be cautious, payoffs or penalties that reward one kind of response more than another, or changes in the frequency of each type of stimuli.

SDT is meant to be used as a tool to measure discriminability and response bias. The likelihood function for the SDT model is easy to compute, which makes it yet another model we can use as a case study to examine other sampling algorithms. In the BCDMEM example above, we examined a single-level model with the ABCDE algorithm. For this example, we will investigate the ability of the Gibbs ABC algorithm to recover both subject-level and hyper-level parameters.

4.3.1 Generating the Data

The parameters for an individual j are that person's discriminability d_j and bias b_j. We built a hierarchy by assuming that each discriminability parameter follows a normal distribution with mean d_μ and standard deviation d_σ, and that each bias parameter follows a normal distribution with mean b_μ and standard deviation b_σ. To generate data to which the model could be fit, we first set $d_\mu = 1$, $b_\mu = 0$, $d_\sigma = 0.20$, and $b_\sigma = 0.05$. We then drew nine d_j and b_j parameters from the normal hyperdistributions defined by the hyperparameters for nine hypothetical subjects.

We used these person-level parameters to generate "yes" responses for $N = 500$ noise and signal trials by sampling from binomial distributions with probabilities equal to the areas under the normal curves to the right of the criterion (i.e., $d_j/2+b_j$).

4.3.2   Recovering the Posteriors

We fit the hierarchical SDT model in two ways. The first way uses the true likelihood function [6, 9]. The second approach used the Gibbs ABC algorithm and a kernel-based ABC algorithm to approximate the likelihood function [55], both of which are described in Chap. 2. We set $\rho(X, Y)$ equal to the Euclidean distance between the observed hit and false alarm rates (i.e., the simulated data to which the model is being fitted) and the hit and false alarm rates arising from simulating the model with a set of proposed parameters θ^*. This distance was weighted with a Gaussian kernel using a turning parameter $\delta = 0.01$. For both the likelihood-free and likelihood-informed estimation procedures, we generated 24 independent chains of 10,000 draws of each parameter, from which we discarded the first 1000 iterations. This left 216,000 samples from each method with which we estimated the joint posterior distributions of each parameter.

4.3.3   Results

Figure 4.2 shows the estimated posterior distributions for the model's hyperparameters, d_μ, b_μ, d_σ, and b_σ, as histograms. Overlaid on these histograms are the posterior density estimates (solid curves) obtained from the likelihood-informed method (i.e., MCMC), and the vertical lines represent the values of the hyperparameters with which the fitted data were simulated. The left panels of Fig. 4.2 show the estimated posteriors for the hypermeans b_μ (top) and d_μ (bottom). The right panels show the estimated posteriors for the log hyper standard deviations b_σ (top) and d_σ (bottom). The estimates obtained from the Gibbs ABC algorithm closely match the estimates obtained using conventional MCMC.

At the individual level, Fig. 4.3 shows the estimated posterior distributions for the discriminability (d_j) parameters. As for the group-level parameters, the two methods produced posterior estimates that do not differ greatly. Although we can also examine the posterior distributions for the subject-level bias parameters (b_j), these estimates were similarly accurate as the discriminability parameters shown in Fig. 4.3.

4.3.4   Summary

We used a combination of Gibbs ABC and kernel-based ABC to estimate the parameters of the SDT model. The likelihood function for this model is well known and simple so the true posterior distribution can be estimated using standard MCMC techniques. We showed that the estimated posteriors using both likelihood-informed

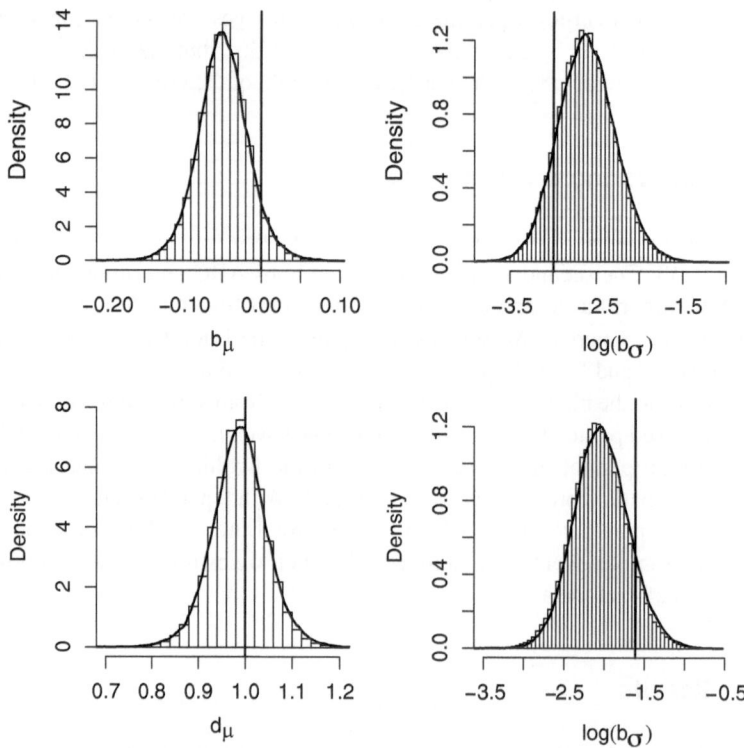

Fig. 4.2 The estimated posterior distributions obtained using likelihood-informed methods (black densities) and the Gibbs ABC algorithm (histograms) for the hyperparameters of the classic SDT model. Vertical lines are placed at the values used to generate the data. The rows correspond to group-level parameters for the bias parameter b (top) and the discriminability parameter d (bottom). The columns correspond to the hypermeans (left) and the hyper standard deviations on the log scale (right)

and likelihood-free methods were similar for both the individual-level and the group-level parameters. These results demonstrate that Gibbs ABC fused with a kernel-based approach can recover the true posterior distributions of the hierarchical SDT model accurately.

4.4   Validation 3: The Linear Ballistic Accumulator Model

The Linear Ballistic Accumulator model (LBA; [95]) is a stochastic accumulator used to explain choice response time data. In a typical choice response paradigm, people are asked to make a decision with two or more response alternatives. For example, in a numerosity task, a person might be asked to select which of two boxes contains more of a certain type of object (e.g., stars, dots, etc.). The time between the onset of the stimulus and the execution of the response is the response time (RT) and the box that is chosen (left or right) determines the choice response. The

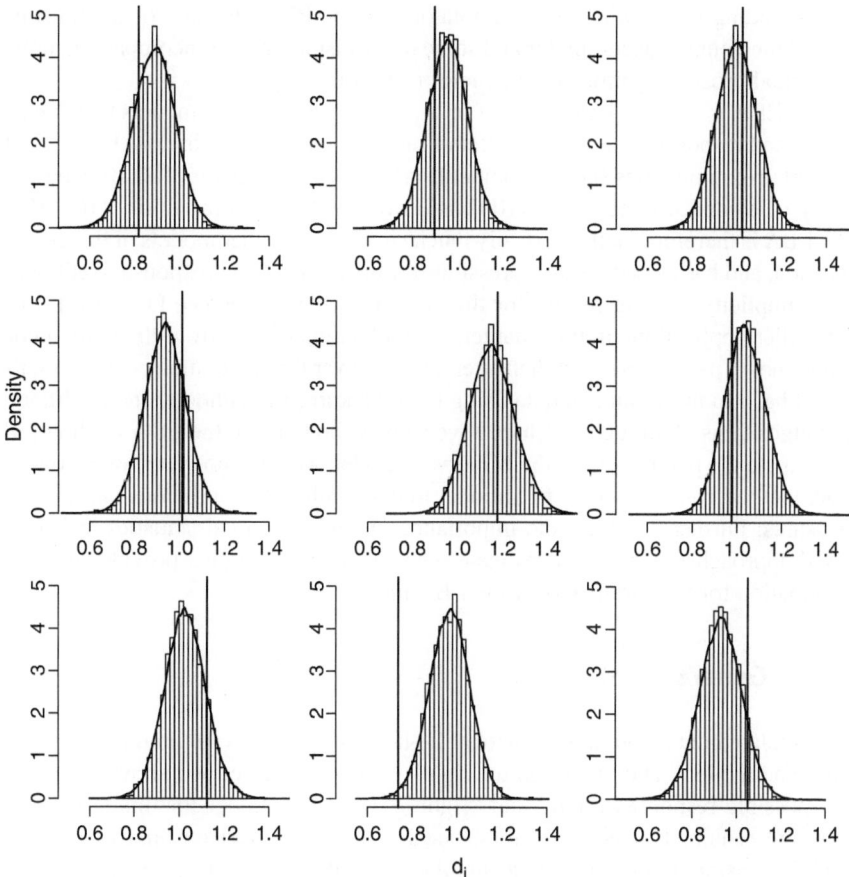

Fig. 4.3 The estimated posterior distributions obtained using likelihood-informed methods (black densities) and the Gibbs ABC algorithm (histograms) for the person-level discriminability parameters d_j of the classic SDT model. Vertical lines are located at the values used to generate the data. Each panel shows the posteriors for a different person

data from this task are therefore mixed, including both continuous RT measures and discrete response measures.

The LBA model postulates the existence of separate accumulators for each possible choice alternative. Each accumulator stores evidence for a particular response, and this evidence increases over time following the presentation of a stimulus. In particular, on accumulator c, the function describing the amount of evidence as a function of time is linear with slope d_c and starting point (y-intercept) k_c. The evidence on the accumulator grows until it reaches a threshold b. The starting point k_c is uniformly distributed in the interval $[0, A]$, and the rate of evidence accumulation d_c is normally distributed with mean $v^{(c)}$ and standard deviation s. After the fastest accumulator reaches the threshold b, a decision is made corresponding to the winning accumulator, and the RT is the sum of the finishing time of the winning accumulator and some nondecision time τ (encompassing initial perceptual processing time and the motor response time).

The LBA is one of a large class of information accumulation models that explain choice and response time data. It differs from other models in this class by eliminating certain complexities such as competition between alternatives [100,101], passive decay of evidence ("leakage") [100], and even within-trial variability [102, 103]. The LBA is therefore mathematically much simpler than other models of this type of decision, and has closed-form expressions for the joint RT and response likelihoods. This simplicity is a big reason for this model's wide acceptance [104–109]. Like the earlier applications in this chapter, we will simulate data from the LBA model using known parameters, and then attempt to recover those parameters by fitting the model back to the simulated data using both the known likelihoods and likelihood-free algorithms. Although in Chap. 3 we showed how to use the PDA method [38] to estimate the parameters of the Minerva 2 model, we have not yet shown how the PDA method can be used to fit a model to data with both continuous and discrete measures. Further, to stress the importance of selecting good statistics in kernel-based approaches, we will investigate whether or not quantiles provide sufficient information for the parameters of the LBA model.

4.4.1 Generating the Data

We simulated data from a hypothetical 2-choice task. For two stimulus types (say, "more on the left" and "more on the right" for the task described above), there are two possible responses ("left" and "right"). The LBA model for this task would therefore require two accumulators, one for the "left" decision and one for the "right" decision. On each trial, accumulator c will have a starting point k_c and an accumulation rate d_c, where c indexes the "left" or "right" accumulator. If a "more on the left" stimulus is presented, the "left" accumulator's rate will be greater on average than the "right" accumulator's rate, and if a "more on the right" stimulus is presented, the "right" accumulator's rate will be greater on average than the "left" accumulator's rate. For simplicity, we assumed no asymmetry between the "left" and "right" accumulators, so we can write the effects of stimulus more compactly in terms of the means of $d_{\text{"left"}}$ and $d_{\text{"right"}}$, letting $v^{(C)}$ represent the mean accumulation rate for correct responses ("left" to "more on the left" and "right" to "more on the right") and $v^{(I)}$ represent the mean accumulation rate for incorrect responses ("right" to "more on the left" and "left" to "more on the right") and setting $v^{(C)} > v^{(I)}$.

We generated 500 responses from the LBA model using a threshold $b = 1.0$, setting the upper bound of the uniform starting point distribution $A = 0.75$, setting the mean accumulation rate for correct responses $v^{(C)} = 2.5$, and the mean accumulation rate for incorrect responses $v^{(I)} = 1.5$. We also added to each simulated RT a nondecision time $\tau = 0.2$. We set the standard deviation of the accumulation rates to $s = 1$ to satisfy the scaling properties of the model. All of these parameter values are consistent with previously published fits of the LBA model to experimental data.

4.4.2   Recovering the Posterior

We used three different approaches to estimate the posterior distribution of the model parameters (i.e., $b, A, v^{(I)}, v^{(C)}, \tau$). The first approach makes use of the likelihood function (see [57, 95, 108, 109], for applications). The second is the PDA method for mixed data types as described in Chaps. 2 and 3. When using this method, we simulated the model 10,000 times for each parameter proposal. The third is the synthetic likelihood algorithm [43], which requires the specification of a set of summary statistics $S(\cdot)$. To implement the algorithm, we decided to use the sample quantiles (corresponding to the cumulative proportions $\{0.1, 0.3, 0.5, 0.7, 0.9\}$) for both the correct and incorrect RT distributions.[1] Thus, for a given set of RTs Y and choices Z, we summarized the data by computing the vector $S(Y, Z)$ comprising 11 statistics: 5 quantiles for each of the samples of correct and incorrect RTs, plus the proportion of correct responses. When using the synthetic likelihood method, we generated 50,000 model simulations per parameter proposal.

It has been noted that the parameters of the LBA model are highly correlated by examining the correlation of samples from the joint distribution of model parameters (see [57]). The correlation in the posteriors makes it difficult to propose sets of parameters that will be accepted at the same time. As a result, conventional sampling algorithms such as Markov chain Monte Carlo (MCMC) [46] can be inefficient, requiring very long chains, and are therefore impractical. For this reason we used the DE-MCMC algorithm to draw samples from the posterior distribution for each of the three methods. For each of the three different likelihood evaluation methods, we implemented a DE-MCMC sampler with 24 chains for 5000 sampling iterations, discarding the first 100 observations in each chain.

4.4.3   Results

Figure 4.4 shows the estimated posterior distributions obtained using the PDA method (top row) and the synthetic likelihood method (bottom row). The columns of Fig. 4.4 correspond to the threshold parameter b, the starting point upper bound parameter A, the accumulation rate for correct responses $v^{(C)}$, the accumulation rate for incorrect responses $v^{(I)}$, and the nondecision time τ. In each panel, the true estimated posterior distribution (i.e., the posterior obtained using the true likelihood) is shown as the black curve plotted over the histogram, and the vertical dashed line marks the value of the parameter used to generate the data.

The figure shows two important things. First, the histograms from the PDA method align closely with the true density (black curve) and are centered at the values of the parameters that generated the data. Therefore, we can state that the PDA method produces posterior estimates that are close to the true posterior

---

[1] We treated the correct and incorrect RT distributions as the accumulators themselves, rather than the response alternatives.

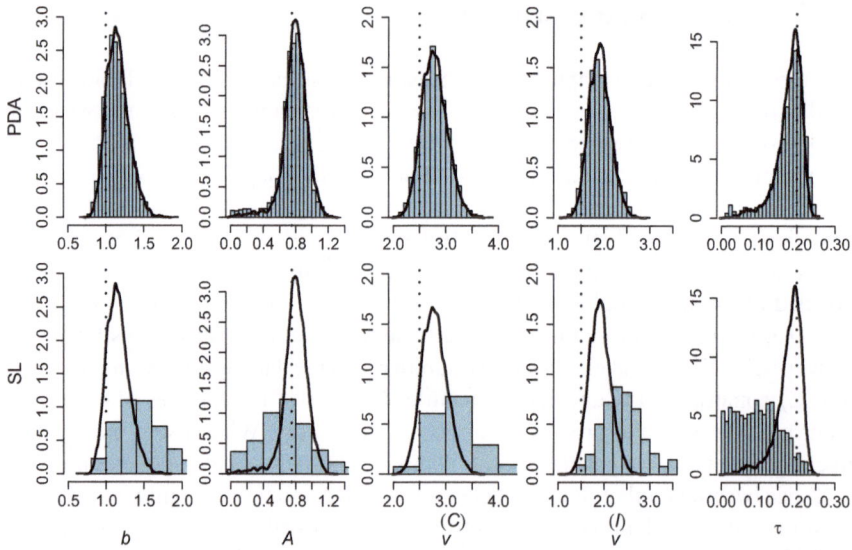

Fig. 4.4 Estimated marginal posterior distributions obtained using the PDA method (top row), and the synthetic likelihood algorithm (SL; bottom row). In each panel, the true estimate of the posterior distribution (i.e., the likelihood-informed estimate) is shown as the black density, and the vertical dashed lines are placed at the values of the parameters used to simulate the data

estimates. Because the PDA method is a general technique that makes use of all the observations in the data set, we can be sure the accuracy of the posterior estimates only depends on the kernel density estimate. Second, the histograms generated using the synthetic likelihood methods vary widely around the true posterior densities and are not centered on the values of the parameters that generated the data. Therefore, we can state that the posterior estimates obtained using the synthetic likelihood method are probably inaccurate. There may be several reasons for this, but the most likely is that the summary statistics (i.e., the quantiles) used for the parameters of the LBA model are not sufficient. The use of quantiles seems to have resulted in high proposal rejection rates even in the high-probability regions of the posteriors.

4.4.4   Summary

In this section we showed that the PDA algorithm can produce accurate estimates of the posterior distributions of the parameters of the LBA model. These results are reassuring, because they imply that the problem of generating sufficient statistics for a model with no implicit likelihood can be safely bypassed with the PDA method. The PDA method does not require the use of sufficient statistics. Using this method we demonstrated accurate recovery of the posterior distribution with only minimal assumptions and specifications of how to approximate the likelihood function. By contrast, the synthetic likelihood algorithm did not produce accurate estimates of the posterior. While there are several reasons why this can happen, we suspect the main reason is that the quantiles we used are not sufficient for the parameters of interest.

This study provides a cautionary tale about the use of likelihood-free algorithms in inappropriate circumstances. Rejection-based and kernel-based algorithms are likely to produce errors in the estimated posterior distribution when sufficient statistics are not known. In these situations, we recommend using the PDA method if computational resources are available to make extensive model simulation feasible.

4.5 Conclusions

In this chapter, we have illustrated the effectiveness of different likelihood-free techniques for three popular models in cognitive science. The models we chose all have a likelihood function, which enabled us to make comparisons between the estimates obtained using likelihood-free methods and the estimates obtained when using the true likelihood function.

In the first application, we showed how the ABCDE algorithm could be used to estimate the posterior distribution of BCDMEM's parameters. We compared the estimates obtained using ABCDE to those of the exact and asymptotic equations put forth by Myung et al. [96]. In the end, we concluded that the estimates obtained using likelihood-free algorithms were not only as accurate as those obtained when using the true likelihood function, but they were achieved at a much faster computation time. We find this promising for the practical implementation of ABCDE, at least for BCDMEM.

In the second application, we showed how a combination of Gibbs ABC and a kernel-based approach could be used to accurately estimate the parameters of a hierarchical SDT model. We found that at both the subject- and group-levels, the estimates obtained using our algorithms were accurate, given that they closely resembled the shape of estimates obtained using a standard, likelihood-informed Bayesian approach.

In the third application, we showed that the PDA method could be used to estimate the parameters of the LBA model. The application of likelihood-free algorithms is tricky when the data consist of choice and response time measures because it is unclear what statistics should be used to convey sufficient information about the full response time distribution to the model parameters. For this reason, we used the PDA method to reconstruct the entire choice RT distribution for a given parameter proposal θ^*, which we then compared to the observed data. We concluded that even in the context where sufficient statistics are not known, if we choose the appropriate likelihood-free algorithm (see Chap. 2), we can still arrive at accurate estimates of the posterior distribution.

Applications

5

5.1 Introduction

In Chap. 4, we applied different likelihood-free algorithms to several different cognitive models. These models all had tractable expressions for the likelihood functions, which meant that we could estimate their true posterior using standard posterior sampling techniques. We used these models to validate our likelihood-free approaches by comparing estimates obtained by fully Bayesian treatments of the problem to those obtained using our approximation methods. After three examples, we concluded that the likelihood-free algorithms we used all provided reasonably good approximations to the true posteriors.

However, it is rarely the case that we would want to approximate the posterior distribution using likelihood-free techniques when the likelihood function is known.[1] For this reason, in this chapter we will highlight a few real-world applications where simulation-based models (with no likelihoods) are fit using likelihood-free techniques. In each case, the model fitting process reveals interesting facts and comparisons that would have otherwise been difficult or impossible to discover.

The outline of this chapter parallels that of Chap. 4. In the first example, we fit the Retrieving Effectively from Memory model [83]. The REM model is also a model of recognition memory, but it makes very different assumptions about the encoding and retrieval processes. In the second example, we fit a dynamic extension of the classic signal detection theory model, which we fit in the previous chapter. This model uses a complex updating process to gradually inform the representations used by observers, whereas in the classic signal detection theory model, the representations are assumed to be fixed throughout the duration of the

---

[1]Sometimes evaluating the likelihood function is more arduous than simply simulating the model, as in our BCDMEM example from Chap. 4.

© Springer International Publishing AG 2018

J.J. Palestro et al., *Likelihood-Free Methods for Cognitive Science*,
Computational Approaches to Cognition and Perception,
https://doi.org/10.1007/978-3-319-72425-6_5

experiment. This model represents an interesting case study—when the model was first developed, we used hand-held fits to obtain good estimates of the parameters via extensive simulations and approximate least squares minimization. However, with the advent of likelihood-free techniques, we can now fit this model to the same data that these hand-held fits were applied, and in turn we can now compare and contrast this Bayesian solution to the approximate least squares one. In the third example, we compare two complex, stochastic accumulator models fit to a simple perceptual decision making task. Prior to this application summarized below, neither of these models had been fit to data in a Bayesian context, which again highlights the importance of likelihood-free techniques across the domain of cognitive science.

5.2   The Retrieving Effectively from Memory Model

In this section, we summarize a recent application of likelihood-free methods to compare models of episodic recognition memory [92]. In this application, we used Gibbs ABC combined with ABCDE to compare the fits of two prominent cognitive models that make different assumptions about how interference occurs in memory: the Bind Cue Decide Model of Episodic Memory [85] and the Retrieving Effectively from Memory (REM) model [83]. These two models make different assumptions about the source of noise in recognition decisions. As discussed in the last chapter, BCDMEM posits that interference from different contexts in which a probe item appeared makes recognition difficult. On the other hand, REM posits that interference from the other items in the study context makes recognition difficult.[2] These two different ideas about interference are built into the structure of the models, leading them to make different predictions about performance under a range of experimental conditions. Both models are very powerful and can fit a range of different experimental effects. However, their relative goodness of fit to data has not been examined in a rigorous way because, much like other simulation-based models, their analytic forms are difficult to derive [96].

Rather than summarizing both model fits to data, we can summarize the hierarchical REM model because at the time of the model comparison analysis in Turner et al. [92], the REM model's likelihood had yet to be derived (but see [110]). Although some variants of the REM model include contextual noise [111], we used the original REM model that does not include this component [83]. As a global memory model, the recognition responses that REM produces are based on a calculation of memory strength arising from a comparison between a probe item and all the items stored in memory. Each stored item is represented as a vector of w features, each of which having some psychological interpretation (such as the extent to which the item "seal" is associated with the concept "wet"). Each feature can take on positive integer values, and the probability that a feature ("wet") takes

---

[2]Although later instantiations of REM incorporate both item and context noise, for our purposes we only consider the pure item-noise version for demonstration.

on a particular value (say, 3) is determined by a geometric distribution, so that the probability that feature K equals value k is given by

$$P(K = k) = g(1 - g)^{k-1} \text{ for } k \in \{1, 2, \ldots, \infty\},$$

where the parameter $g \in (0, 1)$ is called the environmental base rate.

The value of the environmental base rate g is influenced by word frequency. For high-frequency words g is larger than for low-frequency words. Larger values of g will decrease the mean and variance of the feature value K. High-frequency words will therefore tend to have smaller feature values and, because these values take on only positive integer values, have more features in common. This means that high-frequency target words will be more difficult to discriminate from high-frequency distractors.

When an item is studied, its features are copied to a memory trace, but not with complete fidelity. With probability u, a feature will be copied, and with probability $1 - u$, the feature in the trace will remain empty with a value of 0. If a feature is copied, it may not be copied correctly. It will be copied correctly with probability c, and with probability $1 - c$, a random feature value will be sampled from a geometric distribution with parameter g. After all the features of all items are copied into the trace, the result is an "episodic matrix," the dimensions of which are determined by w, the number of features, and the number of items N_{STUDY} on the study list.

A probe item presented for recognition is compared to each trace in the episodic matrix. Following the notation in [83], we let n_{jq} be the number of nonzero mismatching features in the jth trace, and n_{ijm} be the number of nonzero matching features in the jth trace with a value of i. Then, the similarity λ_j of the jth trace is

$$\lambda_j = (1 - c)^{n_{jq}} \prod_{i=1}^{\infty} \left[\frac{c + (1 - c)g(1 - g)^{i-1}}{g(1 - g)^{i-1}} \right]^{n_{ijm}}. \tag{5.1}$$

The overall memory strength Φ of the probe is the average similarity across all the traces to which it was compared, or

$$\Phi = \frac{1}{N_{\text{STUDY}}} \sum_{j=1}^{N_{\text{STUDY}}} \Lambda_j. \tag{5.2}$$

The memory strength Φ is a likelihood ratio (the probability that the probe is a target divided by the probability that the probe is a distractor). This implies that an optimal decision rule is then to respond "old" if $\Phi < 1$ and "new" if $\Phi > 1$.

Turner et al. [92] fit a hierarchical version of the REM model to the data presented in [4], which was presented in Chap. 3. To model the word frequency effects in [4], we assumed a new parameter γ for high-frequency words. For simplicity, we created a bivariate indicator variable W_j, such that if the word presented on trial j is a high-frequency word, $W_j = 1$, and $W_j = 0$ for low-frequency words. The environmental base rate can then be written as $g(1 - W_j) + \gamma W_j$.

To model the effects of a filler task, which, in REM, is not linked to any of the parameters, we wanted to make sure that the number of parameters in REM was the same as the number of parameters in BCDMEM. This constraint controls for the most basic definition of model complexity: models with more parameters are often considered to be more complex than models with fewer parameters. We chose to model the effects of a filler task as the addition of spurious traces to the episodic matrix prior to test. Additional traces cause interference and increase the memory strength of distractor items. This approach added only a single parameter to REM, and so the dimensionality of REM matches that of BCDMEM. Using again a bivariate indicator variable F, where $F_j = 1$ for the filler conditions and $F_j = 0$ for the no filler conditions, we defined a new parameter η which represents the number of spurious features added to the episodic matrix. The number of traces in Condition j can then be written as $N_{\text{STUDY}} + F_j\eta$.

The vector length parameter w is not identifiable [110]. By convention, we set w equal to 20 [26, 83]. For notational convenience, we again define $\theta_{i,j}$ as the vector of parameters for the ith person and jth condition, so

$$\theta_{i,j} = \{N_{\text{STUDY}} + \eta_i F_j, g_i(1 - W_j) + \gamma_i W_j, u_i, c_i, w\}. \tag{5.3}$$

Hence, each subject had five free parameters: c_i, g_i, u_i, γ_i, and η_i. We imposed a simple hierarchal structure such that each of these five parameters had an associated group-level mean and standard deviation parameter to model variation across subjects. We refer the reader to [92] for the specific details of the hierarchical models. For illustration, Fig. 5.1 shows a graphical diagram for this model. There are five lower-level parameters to measure individual-level effects and ten hyperparameters to model the group-level effects. We represent the number of hits for Person i in Condition j as $O_{i,j,T}$ and (similarly) false alarms as $O_{i,j,D}$.

5.2.1 Fitting the REM Model to Data

In both [96] and [110], the expressions defining the likelihood functions for BCDMEM and REM, respectively, require extensive numerical integration that are difficult and time consuming. For example, in Chap. 4 we showed that estimates of the joint posterior distribution for the BCDMEM model could be accurately estimated using the ABCDE algorithm [56], and these estimates were obtained 55 times faster than when using the analytic expressions from Montenegro et al. [96]. Although the analytic expressions are useful for validating the likelihood-free approach, fitting hierarchical versions of these models with likelihood-free methods is a fast and accurate way to approximate the joint posterior distributions.

We used a combination of two algorithms to approximate the likelihoods and to fit the hierarchical REM model. First, we used differential evolution (DE) [57, 60] within ABCDE (see Chap. 2) [56] to generate proposals. Second, we used Gibbs ABC (see Chap. 2) to fit the hierarchical models [70]. Recall that the Gibbs ABC algorithm takes advantage of the fact that the group-level parameters do not depend

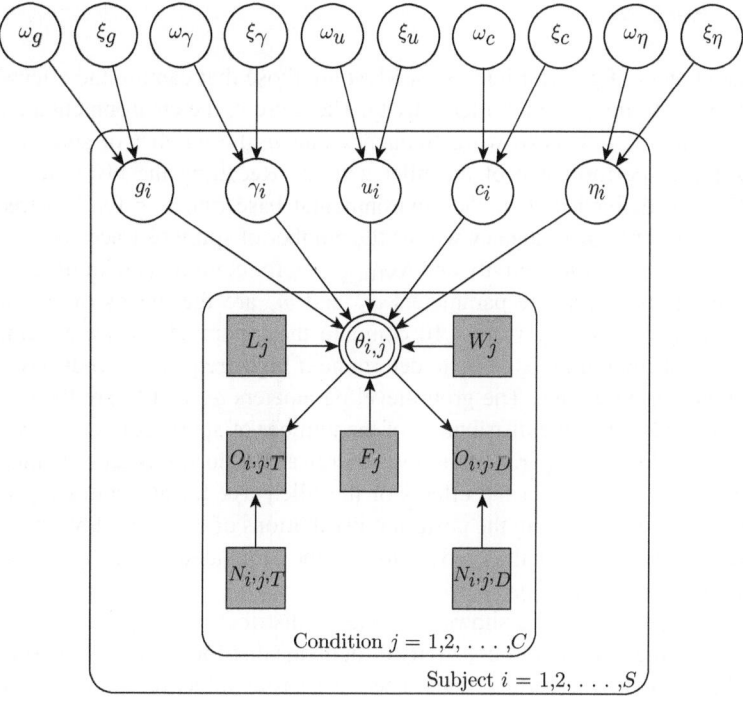

Fig. 5.1 A graphical diagram for the hierarchical REM model for the data of Dennis et al. [4]

on the unknown likelihood function, because the group-level parameters depend on the data only through the individual-level parameters. Thus, if the parameter space is partitioned appropriately, we can obtain samples directly from the posterior distribution of the group-level parameters using a technique called Gibbs sampling and without any need for likelihood-free techniques.

To approximate the likelihood function for each subject, we used a kernel-based approach with normal kernel function such that

$$\mathscr{K}(x|\delta_{\mathrm{ABC}}) = \frac{1}{\sqrt{2\pi}\delta_{\mathrm{ABC}}} \exp\left[-\frac{x^2}{2\delta_{\mathrm{ABC}}^2}\right], \tag{5.4}$$

where $\delta_{\mathrm{ABC}} = 0.2$. Using the distance function $\rho(X, Y) = ||X - Y||$, we evaluated the fitness of the proposal $\theta_{i,j}^*$ by calculating

$$\psi(\theta_{i,j}^* \mid X, Y) = \prod_{i=1}^{S}\prod_{j=1}^{C} \mathscr{K}\left(\rho(\hat{O}_{i,j,T}, O_{i,j,T}) \mid \delta_{\mathrm{ABC}}\right) \mathscr{K}\left(\rho(\hat{O}_{i,j,D}, O_{i,j,D}) \mid \delta_{\mathrm{ABC}}\right), \tag{5.5}$$

where we denote the number of simulated hits as $\hat{O}_{i,j,T}$ and the number of simulated false alarms as $\hat{O}_{i,j,D}$.

5.2.2   Results

The parameters of greatest interest in REM are those that capture the effects of list length, word frequency, and filler activity. These are g, the environmental base rate that explains the effects of word frequency and η, the number of spurious traces added during performance of the filler activity. Recalling the bivariate indicator variables we defined earlier, the environmental base rate is g for low-frequency words and γ for high-frequency words; the number of spurious traces is N_{STUDY} for conditions with no filler activity and $N_{\text{STUDY}} + \eta$ for conditions with filler activity.

At the group level, the parameters ω_γ and ω_g are the means of γ and g for high- and low-frequency words. To compare these parameters, we examined the posterior distribution of $\omega_\gamma - \omega_g$ to determine if high-frequency words have higher environmental base rates. The group-level parameters ω_η and ξ_η are the mean and standard deviation of the distribution of the number of spurious traces, from which the 48 individual-level parameters η were drawn. Because η is an individual-level parameter, we can predict the effects of the filler task for an arbitrary person by simulating values of η from the posterior distributions of the group-level parameters ω_η and ξ_η. We will call the distribution of the simulated values $\tilde{\eta}$ the posterior predictive distribution of η.

The left panel of Fig. 5.2 shows the posterior distribution of $\omega_\gamma - \omega_g$, and the right panel shows the effects of filler activity through the distribution of $\tilde{\eta}$. Zero (no effect) is marked with dashed vertical lines. There is a strong effect for the word frequency manipulation: the probability that $\omega_\gamma > \omega_g$ is 1.0. Therefore, an explanation for how the model accounts for word-frequency effects is that memory traces for low-frequency words are more variable and hence contain more distinctive features than the memory traces for high-frequency words [112]. The difference in the number of distinctive features results in greater discriminability of low-frequency over high-frequency words.

The right panel of Fig. 5.2 shows the distribution of $\tilde{\eta}$. Had there been no effects of the filler activity, the distribution of $\tilde{\eta}$ would be concentrated around zero. However, most of the distribution is well to the right of zero, indicating that most people are sensitive to the filler conditions. The heavy right tail, which extends to around 150, suggests that some people may indeed be sensitive to the filler task. The mode of the distribution is 6, which provides a point estimate of the number of spurious traces we might expect to be added with a filler task.

5.2.3   Concluding Remarks

In this application we demonstrated that ABC provides a way to perform Bayesian inference with the REM model. The results of the analyses provide a way to fit the models to data and interpret their parameters that was not previously possible with standard techniques. Turner et al. [92] were able to directly compare both the BCDMEM and REM models, and concluded that BCDMEM provided a better

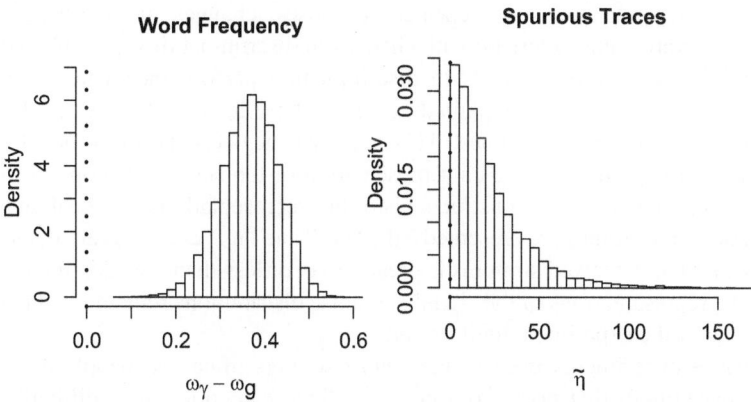

Fig. 5.2 Estimated experimental effects for the word frequency manipulation (left panel) and at the filler manipulation (right panel). The null value of zero is represented by the dashed vertical line in both panels

explanation for some list-length data than did the REM model. Turner et al. concluded that the REM model performed worse because it predicts a list-length effect while not all of the subjects show such an effect.

Had we applied frequentist techniques (such as least squares approximation) to fit these models to data, it would have been difficult to discriminate between them, and we may have come to the conclusion that both models fit the data equally well. The likelihood-free techniques discussed here allowed Turner et al. to exploit a fully hierarchical Bayesian analysis, which allowed for better discrimination between the BCDMEM and REM models. Namely, the REM model incurred a heavy penalty for always predicting a list length effect, when some subjects did not show such effects. Thus, our likelihood-free techniques exposed a critical distinction between the two models that might have been missed by traditional or frequentist modeling techniques.

5.3   A Dynamic Stimulus-Driven Model of Signal Detection

In Chap. 4, we discussed the signal detection theory [71] model, and followed that discussion with a demonstration of accurate parameter recovery using a hierarchical likelihood-free technique. This result, while reassuring, is only a toy problem because the likelihood function for the SDT model is simple and well known. In this section, our focus is on performing likelihood-free inference on an extension of the basic SDT model, a model that we refer to as the dynamic stimulus-driven (DSD) [113] model of signal detection. The DSD model was proposed as an alternative to the basic SDT framework, to accommodate a number of theoretical concerns that we will now briefly review.

As powerful and ubiquitous as the SDT framework has been in psychology, there are several empirical findings for which it cannot account [71]. These include sequential dependencies over repeated responses, changes in the shape of the receiver operating characteristic with changes in discriminability, payoffs and prior probabilities, and improvements in discrimination performance with experience (learning). Most attempts to explain these effects have focused on how people adjust decision criteria with experience. This approach neglects the question of how a person in an experiment builds a representation of the statistical properties of the stimuli (i.e., how a person constructs his or her representations of signal and noise over time as the stimuli are presented); the "yes" and "no" decisions are determined by whether the perceived stimulus is greater than or less than the adapting criterion while the representations of the signal and noise distributions remain fixed from the beginning of the experiment until its end.

Because experiments used in laboratory settings often ask people to classify unfamiliar stimuli that arise from poorly defined categories, it is difficult to see how a person could select an appropriate criterion to guide his or her "yes" and "no" decisions when only a poor representation of these unfamiliar stimuli exists. In developing the DSD model, we were particularly interested in explaining how observers develop stimulus representations that allow them to discriminate between signals and noise, and that could change as the stimulus environment changes. This resulted in a new "dynamic" SDT model that describes how these two distributions are not only constructed, but also evolve over time. The model makes no assumptions about how people set appropriate response criteria. In the next section, we describe the DSD model, and then we use a likelihood-free approach to fit a hierarchical version of the model to empirical data.

5.3.1   Overview of the DSD Model

In this section we will briefly highlight and explain the three main components of the model. The first of these components involves how individuals establish their "priors" about the signal and noise representations. We assume individuals use one of two methods: (1) they either bring a prior, based on past performance, that is useful enough for the discrimination task at hand; or (2) they use the instructions provided by the experimenter during the experiment to develop a minimally informative prior over an impoverished representation of the decision axis.

To illustrate how this may work, consider a person being asked to perform a task that is completely novel and, therefore, he or she has not brought with them a useful prior. The person must construct a minimally informative prior before the presentation of the first stimulus based on the description of the task and the experimenter's instructions. This representation will require both a "support," which is some idea of the range of stimuli that might be experienced, and also an initial likelihood for the two different stimuli (signal and noise) at a few points on that support.

For example, the task that we used asked people to decide if a hypothetical patient is healthy or ill based on a presented "blood assay" value that ranged from 1 to

99. Using this information, which is available to everyone prior to performing the task, someone would construct a rough representation of what a signal and noise trial would look like on the number line between 1 and 99. These representations take the form of locations or points between 1 and 99. Each person might use a different strategy for establishing these representations: some might place points "on the 5s" (5, 10, 15, etc.), some might place points "on the 10s," and so forth. If the experimenter provides additional information by noting that the stimuli will be generated from a random pool of numbers with a mean of 45, then people may construct a prior that places points clustered around 45.

The task requires not only that each person decide *where* to place these representations, but also decide *how many* representation points to use. Unfortunately, humans are notoriously bad at keeping track of large amounts of information over a short period of time, so it is unlikely that people will be able to maintain representations with large number of points. Also, it is possible that the stimulus stream used in the task may change or evolve over time, so people with inflexible representations will not be able to adjust to these changes. For this reason, the DSD model includes a mechanism that allows points within a representation to shift in response to the statistical properties of the stimulus stream.

The second main component of the model is a mechanism for updating priors in response to information available during the task: the currently presented stimulus, the response to that stimulus, and any feedback that may be provided. As the person is exposed to new information during the task, he or she is able to increase (or decrease) their estimate of the likelihood of each representation (signal or noise) based on the magnitude of the currently presented stimulus.

For example, assume that a signal stimulus y_n is presented on trial n. After responding to this stimulus, the observer's signal and noise distributions will be either increased or decreased depending on the feedback she received. If she receives feedback that y_n was a signal (S), then her representation of the S distribution $f_{S,n}$ would increase close to the value y_n, and the representation of the noise (N) distribution $f_{N,n}$ would decrease close to the value y_n. If, on the other hand, she receives feedback that y_n was noise, then the representation of the S distribution $f_{S,n}$ would decrease close to the value y_n, and the representation of the N distribution $f_{N,n}$ would increase close to the value y_n.

Figure 5.3 demonstrates how the S and N representations evolve over time in the DSD framework. The grey dotted lines show the N representations, and the black dotted lines show the S representations. Each dot is a representation point. Early in the experiment (top panel), the representations are sparse, using only a few points. Later on, after more stimuli are presented and more information has been presented available, the representations evolve to look more like the (true) distributions from which the stimuli were generated (solid lines). Finally, after many trials the representations closely resemble the true distributions (bottom panel).

The third and final component specifies how people use the stimulus representations to make decisions. The basis of the decision, as in the classic SDT model, is determined by the likelihood ratio of the presented stimulus. If the estimated likelihood that the stimulus is a signal is greater than the likelihood that it is noise,

Fig. 5.3 An example of how
the DSD model evolves the
representations (dotted lines)
for both signal (black) and
noise (gray) to match the true
stimulus-generating
distributions (solid lines). The
top, middle, and bottom
panels show the DSD model's
representations after 5, 50,
and 100 trials

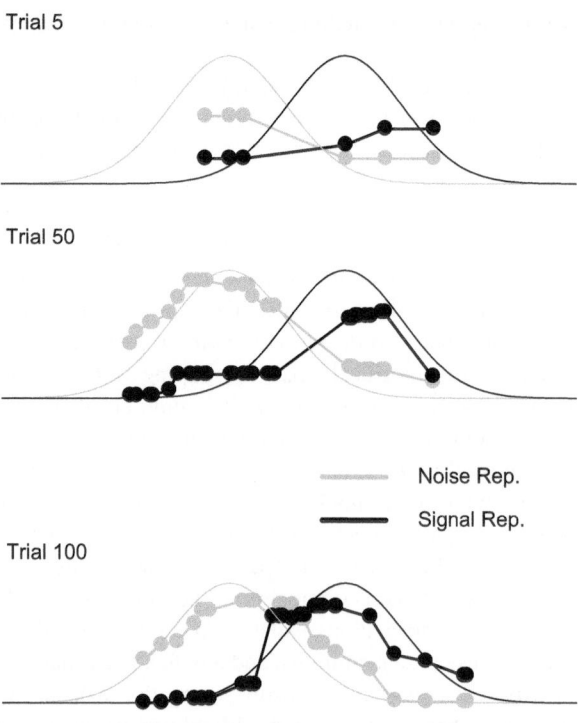

Trial 5

Trial 50

Trial 100

Noise Rep.

Signal Rep.

then the person should respond "yes" [112]. However, in this model, the stimulus representations contain estimates of the likelihoods of signal and noise at only a few representation points. Thus, we assume that on any given trial, the person accesses the point that is closest to the perceived magnitude of the presented stimulus and bases his decision on which likelihood is the highest at this location.

The likelihood ratio decision strategy is equivalent to setting a criterion along the axis of perceived magnitude, as long as the likelihood ratios are monotonic with stimulus magnitude. We might also assume that, depending on task demands, a person may choose not to decide "yes" until the S likelihood of a presented stimulus is not only greater than the N likelihood of that stimulus, but also until the S likelihood exceeds the N likelihood by some amount greater than zero. This strategy would be similar to moving the criterion to the right of the equal-likelihood crossover point, as long as the likelihood ratios are monotonic. In the DSD framework, because the representations are highly dependent on the previously presented stimuli and the responses made to them, there is no reason to believe that the likelihood ratio will be monotonic with perceived stimulus intensity. Therefore, there may be no fixed criterion as in the traditional sense, and there may be more than one location along the axis of perceived magnitude where the likelihood ratio is equal to one.

As useful as the DSD model is in explaining how a person's signal detection strategies change over time, the fact that the stimulus representations evolve over trials means that it is difficult to generate model predictions. It is a simulation-based model with no closed-form likelihood, like the memory models BCDMEM and REM. We will now fit a hierarchical version of this model using the likelihood-free approach, which will permit us to make Bayesian inferences that were not possible in the original study.

5.3.2   The Data

The data we will use in this section comes from the low discriminability condition of Experiment 1 in Turner et al. [113]. Thirty-one people were presented with blood assays from 340 "patients" and were asked to decide whether each patient was healthy or ill. A "yes" response was a decision that a patient was ill, and a "no" response was a decision that a patient was healthy. The blood assay values were drawn from Gaussian distributions with means of 40 for the healthy patients and 60 for the ill patients. Each distribution had a standard deviation of 10. For more details, interested readers should consult Turner et al. [113].

5.3.3   Fitting the DSD Model to Data

The parameters of interest in the DSD model are: γ, the probability of adding/replacing a representation point; λ, the learning rate, which determines how much the representations increase and decrease after a response; δ, the bandwidth, or how far the effect of a stimulus magnitude extends past its representation point; and η, the maximum number of representation points. We assumed that these parameters were allowed to vary freely across subjects. Beyond this, we assumed a mean and standard deviation parameter for each of the four subject-level parameters, so that we could appreciate the group effects.

We used a combination of the Gibbs ABC algorithm and a kernel-based approach (as in the SDT example in the previous chapter) to fit the model. To implement the kernel-based algorithm, we used the Euclidean distance between the observed and simulated hit and false alarm rates for each person, weighted by a Gaussian kernel with a standard deviation of $\delta_{ABC} = 0.01$ to assess the fitness of each proposal. We then estimated the hyperparameters by sampling from their conditional distribution, which do not depend on the approximation of the likelihood function (see Chap. 2). We ran 24 independent chains for 4000 iterations, and treated the first 1000 iterations as burn-in and discarded them. This yielded 72,000 samples with which to form an estimate of the joint posterior distribution of each parameter.

5.3.4   Results

As we've described, the benefit of the likelihood-free Bayesian approach is that it provides substantially more information about a model's ability to fit data in the

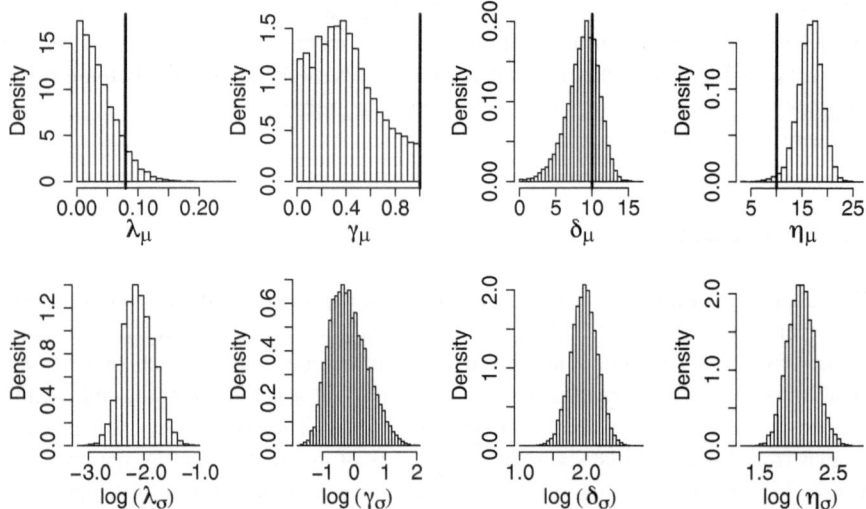

Fig. 5.4 Estimated posterior distributions for each of the hyperparameters in the DSD model. The top row corresponds to the hyper mean parameters whereas the bottom row corresponds to the hyper standard deviation parameters (on the log scale). The left column corresponds to the recency parameter λ, the middle left column corresponds to the probability of representation point replacement γ, the middle right column corresponds to the bandwidth parameter δ, and the right column corresponds to the number of representation points η. The vertical lines in the top row represent the values used by Turner et al. [113] to fit the model to these data, collapsed across individuals

form of parameter estimates at both group and individual levels. To demonstrate this utility, we'll first look at the estimated posterior distributions for each of the parameters shown in Fig. 5.4. These posterior distributions tell us how representations for the signal and noise stimuli are established and maintained.

For example, the posterior of the mean learning rate λ_μ is concentrated on small values (e.g., 0.0–0.10), which suggests that formation of each person's representations is heavily influenced by early experiences with the stimulus set. This means that because (in this experiment) the statistical properties of the stimuli were fixed over trials, it was not necessary to adapt or update the representations after the first few trials.

The posterior estimate for the node replacement probability γ_μ has a mode of approximately 0.4. This small node replacement probability, in combination with a small mean learning rate, is consistent with the idea that people had stable representations that did not vary across trials.

Finally, the posterior estimate of the mean number of representation points η_μ is centered between 15 and 20. The stimuli themselves were drawn from distributions that ranged from 20 to 80, and if we divide this range by the mean of η_μ, we can estimate that people placed representation points approximately 3–4 units apart along the decision axis. Thus, the people in our experiment did not use the full decision axis as assumed by the classic SDT model, but rather made use of a sparse representation.

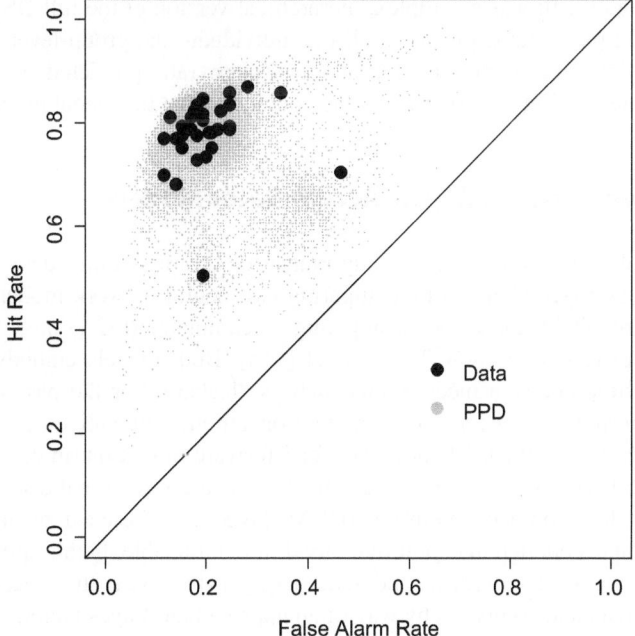

Fig. 5.5 The posterior predictive distribution of the DSD model (gray cloud) along with the data of Turner et al. [113] (black points)

While the estimated posterior distributions provide important information about the model parameters, they do not indicate how well the model fit the data. To examine model fit, we will estimate the posterior predictive distribution of the data, which is obtained by generating predictions from the model conditioned on the sampled parameter estimates. The joint probability density function of the hit and false alarm rates obtained in this way appear in Fig. 5.5 as a gray cloud of points, together with the observed data reported by Turner et al. [113] as black dots. The observed hit and false alarm rates fall neatly within the posterior predictive distribution, providing some assurance that the model fits the data well.

5.3.4.1 Summary and Conclusions

In this section we used Gibbs ABC to estimate the parameters of the dynamic, stimulus driven (DSD) model of signal detection. Unlike the classic SDT model fit in Chap. 3, which has a simple, tractable likelihood function, the DSD has no closed-form likelihood because the representations evolve over time with changes in the stimulus and response stream. With the posterior distributions of the model parameters obtained with these techniques, we can say a lot more about the processes assumed in the model and how they are influenced by experimental variables.

The model we fit was a complex, hierarchical version of the full DSD model. It required 132 parameters, including all the individual- and group-level parameters. We assessed the fit by comparing hits and false alarm rates predicted by the model to those obtained during the experiment and concluded that the model fit the data well.

5.4   Complex, Stochastic Accumulator Models

Turner et al. [114] illustrated the importance of the probability density approximation method (see Chap. 2) by comparing two neural network models of choice response time (RT): the Leaky Competing Accumulator (LCA) model [100] and the Feed-Forward Inhibition (FFI) model [115]. Both models embody neurologically plausible decision mechanisms such as "leakage," or the passive decay of evidence during a decision, and competition among alternatives through either lateral inhibition (in the LCA model) or feed-forward inhibition (in the FFI model). As we will discuss below, these mechanisms extend conventional decision models such as the diffusion decision model (DDM) [103], and these extensions make the models complex enough that their likelihoods are intractable. In this application, we summarize the results in Turner and Sederberg [114] where both models were fit to real-world data and compared by way of an approximate Bayes factor.

5.4.1   The Data

The data to which we will fit the models were collected by Forstmann et al. [106]. Twenty people completed 840 trials across three conditions of a random dot motion task. In this task, people were asked to decide which direction, either left or right, a cloud of dots appeared to be moving. Prior to each trial, people were instructed to respond under one of three conditions: respond quickly (speed direction), respond accurately (accuracy condition), or respond without future instruction (neutral condition). Following each response in the accuracy condition, people were presented with accuracy feedback (correct or incorrect). In the speed and neutral conditions, people were warned if their responses were too slow for the given condition, where "slow" was defined as 400 ms for the speed condition and 750 ms for the neutral condition.

5.4.2   The Models

Turner et al. [114] explored three models. In addition to the LCA and FFI models, they looked at a constrained version of the FFI model. This constrained model performed significantly worse than either the LCA or FFI models, and for this reason, we do not discuss it here.

5.4.2.1 The Leaky Competing Accumulator Model

The LCA [100] model casts a decision as a race between evidence accumulation mechanisms. It was presented as a neurologically plausible model of perceptual decision making. Each accumulator in the race represents a possible response and is modeled as a Gaussian diffusion process with a reflecting boundary at 0. The accumulator that exceeds a threshold level of evidence first wins the race and determines both the response and the RT. For neural plausibility, each accumulator "leaks" evidence, and there is lateral inhibition between the accumulators, so that accumulators with high levels of evidence suppress the accumulation of evidence on accumulators with low levels of evidence.

Panel A in Fig. 5.6 shows the structure of the LCA model for a two-choice decision task. The bottom nodes represent the stimulus, and these nodes are connected to an internal "belief state" (i.e., middle nodes) by the drift rates ρ. It is in the internal belief state where the evidence accumulation process occurs. There is a competition between the response alternatives in the belief state that depends on both the evidence that has been accumulated and lateral inhibition. As the trial progresses and more evidence is accumulated, the alternative with the

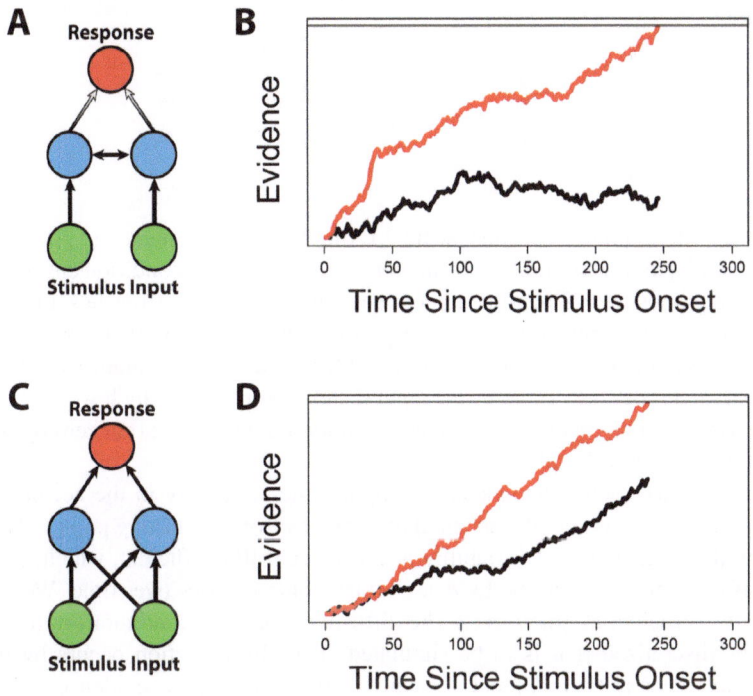

Fig. 5.6 Graphical depiction of the LCA and FFI models. Panels **A** and **B** correspond to the LCA model, whereas Panels **C** and **D** correspond to the FFI model. Panels **A** and **C** show how the stimulus input is mapped to the behavioral response. Panels **B** and **D** show representative simulations of each model in a two-alternative decision task

greatest evidence will inhibit the accumulation of evidence for the other alternative. This inhibition both decreases the evidence on for the other alternative and increases the accumulation of evidence on the alternative with the greatest evidence.

In addition to lateral inhibition, both accumulators lose some of their accumulated evidence to leakage. The parameter κ is the rate of leakage, which depends on the current amount of evidence accumulated. As more evidence is accumulated, more and more evidence will be lost. Eventually, the winning alternative will accumulate enough evidence to cross some threshold α, and a response will be made.

The interaction between evidence, inhibition and leakage is shown in Panel B of Fig. 5.6. At the beginning of a trial when the stimulus is presented, each accumulator starts with the same amount of evidence. Over time, as more evidence is accumulated, one accumulator will gain an advantage. This will result in an inhibition of the other accumulator, such that its level of evidence is reduced toward the starting point. The accumulator in the lead will then accumulate evidence at a faster rater and so it reaches the threshold more quickly.

We denote the rate of evidence accumulation for the cth accumulator as ρ_c, the lateral inhibition parameter as β, and the leakage parameter as κ. The change in the evidence level x_c of the cth accumulator is represented by the stochastic differential equation

$$
dx_c = \left(\rho_c - \kappa x_c - \beta \sum_{j \neq c} x_j \right) \frac{dt}{\Delta_t} + \xi_t \sqrt{\frac{dt}{\Delta_t}}
$$

where Δ_t determines the time scale of the RT measurements (e.g., seconds, milliseconds) and we assume $x_c \leftarrow \max(x_c, 0)$—a boundary condition ensuring that x_c is always positive. When the evidence for any accumulator reaches a threshold α, the process is terminated, and the response corresponding to that accumulator is made. Like many other models of choice RT, we can also include a non-decision time parameter that incorporates perceptual and motor times, which we will denote τ. We assumed that on each trial the accumulation dynamics start at zero by setting $x_c = 0$ for both $c = \{1, 2\}$.

We must also define parameters to capture the variability in the accumulation process, or how smoothly the accumulation moves from its starting point of 0 to the threshold α. The extent of variability is called the drift coefficient, which, because the diffusion processes in the LCA are nonstationary, varies over time. We denote the drift coefficient at time t as ξ_t. The diffusion process is continuous in time, and must be discretized if it is to be simulated. This discretization occurs by taking very small steps in time Δ_t, and at each step incrementing the evidence level by a small amount sampled from a normal distribution with a mean of zero and standard deviation η. In other words, at each time step t in the evidence accumulation process $\xi_t \sim \mathcal{N}(0, \eta)$.

Turner et al. [114] did not employ a hierarchical model. Instead, they fit the LCA and the FFI to each individual independently. They also constrained the drift rate parameters so that $\sum_c \rho^{(c)} = 1$ for each person. The parameters dt (with the unit of seconds) and Δ_t were fixed to 0.01 and 0.1, respectively. Here we will also not fit a full hierarchical model, and we will restrict our examinations to the individual-level parameters.

To capture the effects of the task instruction condition, we assumed three different threshold parameters, creating a total of eight model parameters. At the level of each individual, we specified uninformative priors across all subject-level parameters.

5.4.2.2 The Feed-Forward Inhibition Model

The FFI model differs from the LCA model in two ways: (1) there is no leakage of evidence from the individual accumulators, and (2) lateral inhibition or competition between the accumulators is modeled differently.

Panel C in Fig. 5.6 shows the evidence accumulation process in the FFI model. Like the LCA model, the levels of evidence on the alternatives in the belief state are regulated by the rate of evidence accumulation ρ. However, whereas competition in the LCA model depends on the amount of accumulated evidence for each alternative, the competitive mechanisms in the FFI model depend on the stimulus. This happens by way of a feed-forward inhibition process (the arrows in the figure) regulated by a parameter ν.

Panel D in Fig. 5.6 shows a simulation of the FFI model. When a stimulus is presented, evidence for each alternative begins to accumulate towards a threshold. Competition occurs at the input stage, so gaining an advantage in terms of the amount of evidence accumulated does not increase an accumulator's chances of winning. Inhibition between accumulators depends on the average *input* to other alternatives such that

$$dx_c = \left(\rho_c - \frac{\nu}{C-1} \sum_{j \neq c} \rho_j \right) \frac{dt}{\Delta_t} + \xi_t \sqrt{\frac{dt}{\Delta_t}}$$

where ν is the feed-forward inhibition parameter, ρ_c represents the rate of evidence accumulation for the cth alternative, $\xi_t \sim \mathcal{N}(0, \eta)$ represents the drift coefficient, and $C = 2$ is the number of choice alternatives. Like the LCA, the boundary constraint $x_c \leftarrow \max(x_c, 0)$ prevents the evidence level from becoming negative by reflecting x_c at 0.

As in the LCA above, we assumed that the effects of task instruction could be captured by way of threshold changes. We again constrained the drift rates to sum to one, fixed $dt = 0.01$ (the time scale was measured in seconds), and the time constant parameter $\Delta_t = 0.1$. We also set $x_c = 0$ at $t = 0$ for both alternatives. Hence, the FFI model investigated contained a total of seven parameters.

5.4.2.3 Estimating the Posterior

The LCA and FFI models explain both RT and choice data. Furthermore, the parameters of choice RT models can sometimes be highly correlated [57], which makes conventional sampling algorithms such as Markov chain Monte Carlo inefficient. Therefore, for this analysis, we used the PDA method for mixed data types [38], and DE-MCMC [56, 57, 60] to generate proposals. We implemented the DE-MCMC sampler with 50 chains for 2000 sampling iterations following 500 burn-in iterations, producing 100,000 samples of the joint posterior distributions for each model.

For each proposed parameter, we simulated the models $J = 50,000$ times and performed a log transform on the RTs to obtain stable kernel density estimates. The bandwidth parameters h were calculated for each proposal by way of Eq. (2.25). We then multiplied each density estimate by the proportion of responses appropriate to that estimate to obtain the joint likelihoods of RT and responses under the proposed parameter values.

5.4.3   Comparing the Models

In the context of likelihood-free estimation, the question of how to fairly compare model performance remains elusive [34, 49, 116]. The issue is that while a statistic may be sufficient for all of the parameters from both of the models under comparison, the same statistics may not be sufficient for parameters that characterize model performance [117]. So, because we can't compute the Bayes factor by parameterizing model choice, we must resort to conventional model fit statistics based on the output from our DE-MCMC algorithm. In Turner and Sederberg [114], we used four different metrics to compare the relative fits of the LCA and FFI models: the Akaike information criterion (AIC) [118], the Bayesian information criterion (BIC) [119], the Bayesian predictive information criterion (BPIC) [120], and the Bayes factor. However, for our purposes here, we will only discuss the BIC and Bayes factors (which use the BIC to form an approximation) here. The BIC is obtained by calculating

$$\mathrm{BIC} = -2\log(L(\widehat{\theta}|D)) + \log(N)p, \qquad (5.6)$$

where N represents the number of data points. To obtain a Bayes factor between two model candidates M_q and M_r, given data D, we compute the marginal likelihood of the data under model M_q divided by the marginal likelihood of the data under model M_r, or

$$\mathrm{BF}_{q,r} = \frac{p(D|M_q)}{p(D|M_r)}.$$

There are many other resources discussing how the Bayes factor can be calculated [45, 121]. For our purposes it is sufficient to note that the computation of the Bayes factor is performed by obtaining the marginal likelihoods for each model over all possible values of the models' parameters. This computation, while not complicated, can sometimes encounter difficulties. First, for complex models (such as the FFI and LCA models), the likelihood function is not analytically tractable and must be estimated using numerical integration or approximated asymptotically. Second, because the likelihoods for both the FFI and LCA models are unavailable, we must approximate them. For this application we will use a method presented in Kass and Raftery [122], who showed that, when comparing Models q and r, the difference in BIC values $\text{BIC}_q - \text{BIC}_r$ tends to $-2\log(\text{BF}_{q,r})$ as the sample size increases. Therefore, we can approximate the Bayes factor by evaluating

$$\text{BF}_{q,r} \approx \exp\left[-\frac{1}{2}\left(\text{BIC}_q - \text{BIC}_r\right)\right]. \tag{5.7}$$

5.4.4   Results

We contrasted the fits of the two models by computing the Bayes factor. Figure 5.7 shows the ranked log Bayes factors computed for each person in the experiment. As the log Bayes factors increase, the evidence for the FFI model over the LCA becomes stronger. The dashed black horizontal line at zero is the point where the degree of evidence favors neither model. Subjects with log Bayes factors above zero show stronger evidence for the FFI model, while those with Bayes factors below zero show stronger evidence for the LCA model. Figure 5.7 shows that the magnitudes of the differences from zero are much greater for the log Bayes factors below zero: When the LCA model is the better model, the evidence for it is much greater than when the FFI is the better model.

5.4.5   Conclusion

In this application, we used the probability density approximation (PDA) method to fit two neural network models to the data presented in Forstmann et al. [106]. The first model, the Leaky Competing Accumulator [100] uses neurally plausible mechanisms such as competition via lateral inhibition and leakage. The second model, the Feed-forward Inhibition [115] model, assumes that competition between alternatives follows a feed-forward inhibition process, and it assumes that leakage is not present in the network. Both models are neurally inspired and have been shown to account for a number of experimental effects [100, 115, 123–128].

In Turner and Sederberg [114], neither the AIC nor the BIC measures provided strong evidence that one model should be preferred to the other. However, the BPIC measure favored the LCA model for most (15 out of 20) people. The discrepancies among these different metrics is not surprising. They are only crude measures of

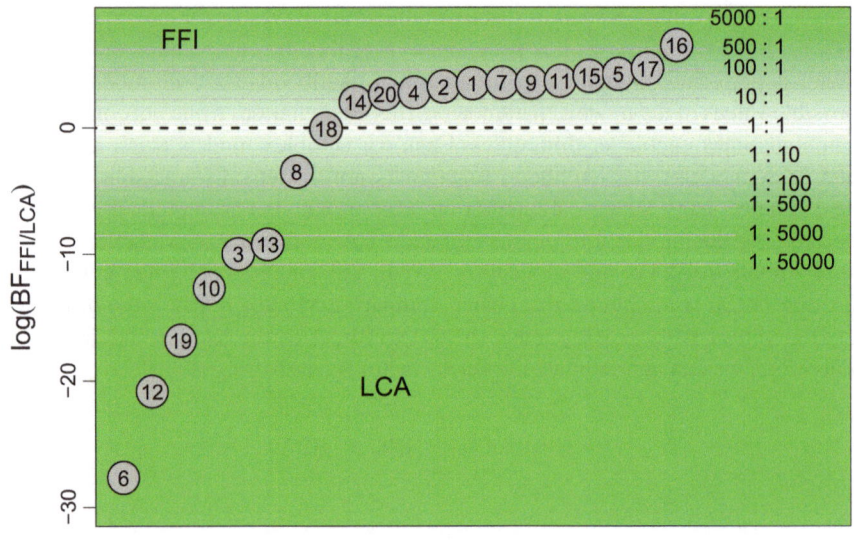

Fig. 5.7 A comparison of the Bayes factors comparing the FFI model to the LCA model for each person. Individuals have been ranked by log Bayes factor, where a higher Bayes factor corresponds to greater evidence for the FFI model. The point of indifference between the two models is represented as the dashed horizontal line at zero

model fit, and more extensive analyses are needed to contrast complex models such as LCA and FFI. We computed an approximation to Bayes factor using Eq. (5.7) [119, 122]. The Bayes factor indicated that the FFI was the better model for 12 of the 20 people in the experiment. However, when the LCA model outperformed the FFI model, the evidence in its favor was much greater than the evidence in favor of the FFI. This finding might suggest that different people in the experiment used different kinds of decision processes. For example, some people might be more prone to leakage of information from the alternatives in the belief state, or that the lateral inhibition between the alternatives was stronger, inducing more temporal dependencies between the alternatives. The explanation for the results could be theoretically interesting, or perhaps the simplifying assumptions we used to fit the models influenced their fits to the data, or that the data are not sufficient to tell these models apart.

Conclusions

This book has provided an overview of several prominent likelihood-free algorithms that are readily available for use when fitting psychological models to data. We began in Chap. 1 by introducing the concepts of approximate least squares, Bayesian inference, and approximate Bayesian computation (ABC). Next, in Chap. 2, we presented several likelihood-free algorithms, along with code for implementing them. Of course there are now many more algorithms available that we could have discussed, but these algorithms are often more complicated and have not yet been used in cognitive science. As such, we limited our focus to algorithms that would be discussed in later chapters. In Chap. 3, we provided a tutorial on fitting the Minerva 2 model to simulated data, and we compared the relative merits of the probability density approximation (PDA; [38]) method, a kernel-based ABC algorithm, and asymptotic expressions for an approximate likelihood function. We then showed how one could extend the kernel-based algorithm hierarchically, and applied it to fit the Minerva 2 model to the data from [4].

Chapters 4 and 5 focused on a small set of recovery and model-fit exercises we have completed in our own research. In Chap. 4, we presented "validations" where each cognitive model had an explicit likelihood function, and so the true posterior could be estimated. We used the true posterior as a metric for evaluating the accuracy of posteriors obtained using various likelihood-free algorithms. These exercises were useful in the applied setting because we have used these examples to provide assurance that we had made appropriate choices when moving to similar models with intractable likelihood functions. Whereas Chap. 4 focused on validating the likelihood-free approach, Chap. 5 provided some interesting examples of fitting models to data whose likelihood functions are currently intractable. Here, we provided summaries of applications in our own work; specifically, we fit the retrieving actively from memory model [44, 83, 92], the dynamic signal detection model [70, 113], the feed-forward inhibition model [114, 115], and the leaky

© Springer International Publishing AG 2018

J.J. Palestro et al., *Likelihood-Free Methods for Cognitive Science*,
Computational Approaches to Cognition and Perception,
https://doi.org/10.1007/978-3-319-72425-6_6

competing accumulator model [100, 114]. While we cannot say whether or not the estimated posteriors in these applications were accurate, we believe them to be based on the simulation results from the "validation" studies.

Our hope is that the techniques discussed in this book may serve as the catalyst in the advancement of mechanistic models of cognition. To us, the primary advantage of likelihood-free techniques is the infinite number of possibilities for new model mechanisms, distributional assumptions, or processing stages. The techniques described here allow freedom from the burden of simplifying assumptions in an effort to acquire mathematical tractability. While it is not clear how often simplifying assumptions are made for the purposes of mathematical tractability, the advantages and disadvantages of the commitment to tractability motivates a stimulating discussion on the role that tractability should play in model development. In developing mathematical models, our goal is to put forth a model that can not only fit data well, but also makes a strong yet accurate commitment to the distribution of data we should see in our experiments [128, 129]. The assessment of a model's full credentials involves two important considerations: model fit and model complexity [130–132].

In the domain of model development, the word complexity can sometimes refer to either the flexibility of a model and can sometimes refer to the ease of implementation [133]. However, these are two different concepts. The ease of implementation is related to the mathematical tractability, but it is not related to complexity [96, 110]. Within this book, we have described several mathematical models that are easy to implement and fit to data because there are analytic expressions relating the model parameters to the data (e.g., signal detection theory). These expressions make the model very easy to fit to data via maximum likelihood or Bayesian approaches. Unfortunately, tractability does not necessarily map onto fewer parameters, or the degree of model flexibility. As such, tractability is also unrelated to complexity when used as a measure of model performance.

To illustrate, consider as analogy the bind cue decide model of episodic memory (BCDMEM; [85]). The BCDMEM model was proposed as a pure context model of episodic memory, an assumption that was at odds with the dominant models at the time. The model was presented as a simulation model, meaning that the likelihood function relating model parameters to predictions about the hit and false alarm rates was intractable. For years, anytime a researcher wanted to fit BCDMEM to their data, they were forced to rely on simulation methods, such as approximate least squares. Eventually, Myung et al. [96] produced analytic expressions for the model. While these expressions are computationally difficult to evaluate, they can be used to assess the model's flexibility, complexity, and identifiability [92]. Ultimately, the expressions derived by Myung et al. [96] unlocked one key facilitator in the endeavor of rigorous model evaluation: mathematical tractability.

What can we make of the research conducted in the time between the development of the original model in 2001, and the derivation of analytic expressions in 2007? As the assumptions of BCDMEM were never changed during this time period, the complexity of BCDMEM also never changed. Hence, the ability of BCDMEM to fit data also never changed. In a similar vein, if a researcher published a paper deriving analytic expressions for say, the complex leaky competing accumulator (LCA; [100]) model tomorrow, nothing about the previous *fits* of the LCA model over the past decade will have changed. Nothing about the model's *complexity* will have changed either. Instead, the LCA model would simply be given a compelling *pragmatic* advantage in choosing among the various models for application purposes because the model would now be (potentially) easier to fit to data (but note the simulation performance differences for the BCDMEM in Chap. 4).

While mathematical tractability is highly advantageous, the analyses in this book highlight the importance of methods for performing inference on simulation-based models. In theory, any computational model can now be fit to data using the likelihood-free approach, allowing researchers to regain access to tried-and-true methods for model evaluation. Our view is that, by using these methods, researchers are free to experiment with as many complex and stochastic model variants as they can imagine, while still assessing model flexibility relative to the data. Of course, tractable models offer compelling advantages, but if compromising assumptions are required to produce tractability, these assumptions may now be rejected on the basis of a theoretical position, prior research, or even curiosity.

Distributions

Here, we provide the PDFs for several distributions that we use throughout the book.

Beta Distribution The probability of observing the random variable x under the Beta distribution with shape parameters $\alpha \in (0, \infty)$ and $\beta \in (0, \infty)$ is

$$f(x|\alpha, \beta) = \frac{\Gamma(\alpha)\Gamma(\beta)}{\Gamma(\alpha + \beta)} x^{\alpha-1}(1 - x)^{\beta-1}$$

where $\Gamma(x) = (x - 1)!$.

Binomial Distribution In n trials, the binomial distribution defines the probability of observing $x = \{0, 1, \ldots, n\}$ successes as

$$f(x|p, n) = \binom{n}{x} p^x (1 - p)^{n-x},$$

where the probability of a single-trial success is the parameter $p \in [0, 1]$.

Gamma Distribution The probability of observing the random variable x under the Gamma distribution with shape parameter $k \in (0, \infty)$ and scale parameter $\theta \in (0, \infty)$ is

$$f(x|k, \theta) = \frac{1}{\Gamma(k)\theta^k} x^{k-1} \exp\left(-\frac{x}{\theta}\right).$$

© Springer International Publishing AG 2018   119
J.J. Palestro et al., *Likelihood-Free Methods for Cognitive Science*,
Computational Approaches to Cognition and Perception,
https://doi.org/10.1007/978-3-319-72425-6_7

References

1. S. Lewandowsky, S. Farrell, *Computational Modeling in Cognition: Principles and Practice* (SAGE Publications, Thousand Oaks, CA, 2010)
2. W.H. Batchelder, D.M. Riefer, Psychol. Rev. **97**, 548 (1990)
3. W.H. Batchelder, D.M. Riefer, Psychon. Bull. Rev. **6**, 57 (1999)
4. S. Dennis, M. Lee, A. Kinnell, J. Math. Psychol. **59**, 361 (2008)
5. M.D. Lee, J. Math. Psychol. **48**, 310 (2004)
6. M.D. Lee, Psychon. Bull. Rev. **15**, 1 (2008)
7. M.D. Lee, J. Math. Psychol. **55**, 1 (2011)
8. M.D. Lee, I.G. Fuss, D.J. Navarro, in *Advances in Neural Information Processing*, 19th edn., ed. by B. Scholkopf, J. Platt, T. Hoffman (MIT Press, Cambridge, MA, 2006), pp. 809–815
9. J.N. Rouder, J. Lu, Psychon. Bull. Rev. **12**, 573 (2005)
10. J.N. Rouder, J. Lu, P. Speckman, D. Sun, Y. Jiang, Psychon. Bull. Rev. **12**, 195 (2005)
11. J.N. Rouder, D. Sun, P. Speckman, J. Lu, D. Zhou, Psychometrika **68**, 589 (2003)
12. R.M. Shiffrin, M.D. Lee, W. Kim, E.J. Wagenmakers, Cogn. Sci. **32**, 1248 (2008)
13. M. Steyvers, M.D. Lee, E.J. Wagenmakers, J. Math. Psychol. **53**, 168 (2009)
14. E.J. Wagenmakers, Psychon. Bull. Rev. **14**, 779 (2007)
15. M.D. Lee, W. Vanpaemel, Cogn. Sci. **32**, 1403 (2008)
16. P. Craigmile, M. Peruggia, T.V. Zandt, Psychometrika **75**, 613 (2010)
17. M.D. Lee, E.J. Wagenmakers, *Bayesian Modeling for Cognitive Science: A Practical Course* (Cambridge University Press, Cambridge, 2013)
18. Z. Oravecz, F. Tuerlinckx, J. Vandekerckhove, Psychometrika **74**, 395 (2009)
19. J. Vandekerckhove, F. Tuerlinckx, M.D. Lee, Psychol. Methods **16**, 44 (2011)
20. J.K. Kruschke, *Doing Bayesian Data Analysis: A Tutorial with R and BUGS* (Academic, Burlington, MA, 2011)
21. I. Klugkist, O. Laudy, H. Hoijtink, Psychol. Methods **15**, 281 (2010)
22. P.S. Laplace, Stat. Sci. **1**(3), 364 (1774/1986). http://dx.doi.org/10.1214/ss/1177013621
23. J.O. Berger, J.M. Bernardo, D. Sun, Bayesian Anal. **10**(1), 189 (2015). http://dx.doi.org/10.1214/14-BA915
24. M.W. Howard, M.J. Kahana, J. Math. Psychol. **46**, 269 (2002)
25. P.B. Sederberg, M.W. Howard, M.J. Kahana, Psychol. Rev. **115**, 893 (2008)
26. A.H. Criss, J.L. McClelland, J. Mem. Lang. **55**, 447 (2006)
27. K.J. Malmberg, R. Zeelenberg, R. Shiffrin, J. Exp. Psychol. Learn. Mem. Cogn. **30**, 540 (2004)
28. J.K. Pritchard, M.T. Seielstad, A. Perez-Lezaun, M.W. Feldman, Mol. Biol. Evol. **16**, 1791 (1999)
29. M.A. Beaumont, Annu. Rev. Ecol. Evol. Syst. **41**, 379 (2010)
30. M.J. Hickerson, E.A. Stahl, H.A. Lessios, Evolution **60**, 2435 (2006)
31. M.J. Hickerson, C. Meyer, BMC Evol. Biol. **8**, 322 (2008)

© Springer International Publishing AG 2018   121
J.J. Palestro et al., *Likelihood-Free Methods for Cognitive Science*,
Computational Approaches to Cognition and Perception,
https://doi.org/10.1007/978-3-319-72425-6

32. V.C. Sousa, M. Fritz, M.A. Beaumont, L. Chikhi, Genetics **181**, 1507 (2009)
33. M.G.B. Blum, O. François, Stat. Comput. **20**, 63 (2010)
34. C. Leuenberger, D. Wegmann, Genetics **184**, 243 (2010)
35. D. Wegmann, C. Leuenberger, L. Excoffier, Genetics **182**, 1207 (2009)
36. D.L. Hintzman, Behav. Res. Methods Instrum. Comput. **16**, 96–101 (1984)
37. C.F. Sheu, J. Math. Psychol. **36**, 592 (1992)
38. B.M. Turner, P.B. Sederberg, Psychon. Bull. Rev. **21**, 227 (2014)
39. B.W. Silverman, *Density Estimation for Statistics and Data Analysis* (Chapman & Hall, London, 1986)
40. W.R. Holmes, J. Math. Psychol. **68**, 13 (2015)
41. J.A. Rice, *Mathematical Statistics and Data Analysis* (Duxbury Press, Belmont, CA, 2007)
42. P. Fearnhead, D. Prangle, J. R. Stat. Soc. Ser. B **74**, 419 (2012)
43. S. Wood, Nature **466**, 1102 (2010)
44. B.M. Turner, T. Van Zandt, J. Math. Psychol. **56**, 69 (2012)
45. A. Gelman, J.B. Carlin, H.S. Stern, D.B. Rubin, *Bayesian Data Analysis* (Chapman and Hall, New York, NY, 2004)
46. C.P. Robert, G. Casella, *Monte Carlo Statistical Methods* (Springer, New York, NY, 2004)
47. S. Sisson, Y. Fan, M.M. Tanaka, Proc. Natl. Acad. Sci. USA **104**, 1760 (2007)
48. P. Del Moral, A. Doucet, A. Jasra, J. R. Stat. Soc. B **68**, 411 (2006)
49. T. Toni, D. Welch, N. Strelkowa, A. Ipsen, M.P. Stumpf, J. R. Soc. Interface **6**, 187 (2009)
50. M.A. Beaumont, J.M. Cornuet, J.M. Marin, C.P. Robert, Biometrika **asp052**, 1 (2009)
51. S. Kullback, J.C. Keegel, J.H. Kullback, *Topics in Statistical Information Theory*. Lecture Notes in Statistics, vol. 42 (Springer, New York, 1987)
52. R. Douc, A. Guillin, J.M. Marin, C. Robert, Ann. Stat. **35**, 420 (2007)
53. M.A. Beaumont, W. Zhang, D.J. Balding, Genetics **162**, 2025 (2002)
54. N.J.R. Fagundes, N. Ray, M. Beaumont, S. Neuenschwander, F.M. Salzano, S.L. Bonatto, L. Excoffier, Proc. Natl. Acad. Sci. **104**, 17614 (2007)
55. R.D. Wilkinson, Biometrika **96**, 983 (2008)
56. B.M. Turner, P.B. Sederberg, J. Math. Psychol. **56**, 375 (2012)
57. B.M. Turner, P.B. Sederberg, S. Brown, M. Steyvers, Psychol. Methods **18**, 368 (2013)
58. J.A. Vrugt, C.J.F. ter Braak, C.G.H. Diks, B.A. Robinson, J.M. Hyman, D. Higdon, Int. J. Nonlinear Sci. Numer. Simul. **10**, 273–290 (2009)
59. B. Hu, K.W. Tsui, Technical Report Number 1112 (2005)
60. C.J.F. ter Braak, Stat. Comput. **16**, 239 (2006)
61. R. Storn, K. Price, J. Glob. Optim. **11**, 341 (1997)
62. G. Tong, Z. Fang, X. Xu, *Evolutionary Computation* (2006), pp. 438–442
63. R. Havangi, M. Nekoui, M. Teshnehlab, Int. J. Comput. Sci. Issues **7**, 15 (2010)
64. J. Kennedy, R. Eberhart, Proc. IEEE Int. Conf. Neural Netw. **4**, 1942 (1995)
65. R. Tanese, Distributed genetic algorithms, in *Proceedings of the Third International Conference on Genetic Algorithms and Their Applications,* ed. by D. Schaffer (Morgan Kaufmann, San Mateo, 1989), pp. 434–439
66. V.A. Epanechnikov, Theory Probab. Appl. **14**, 153 (1969)
67. P. Kontkanen, P. Myllymäki, in *Proceedings of the 11th International Conference on Artificial Intelligence and Statistics* (Artificial Intelligence and Statistics, San Juan, Puerto Rico, 2007)
68. F. Chapeau-Blondeau, D. Rousseau, Physica A **388**, 3969 (2009)
69. L. Excoffer, A. Estoup, J.M. Cornuet, Genetics **169**, 1727 (2005)
70. B.M. Turner, T. Van Zandt, Psychometrika **79**, 185 (2014)
71. D.M. Green, J.A. Swets, *Signal Detection Theory and Psychophysics* (Wiley, New York, 1966)
72. J.P. Egan, Recognition memory and the operating characteristic. Technical Report, AFCRC-TN-58-51, Hearing and Communication Laboratory, Indiana University, Bloomington, IN (1958)
73. B.B. Murdock, Psychol. Rev. **89**, 609 (1982)
74. G. Gillund, R.M. Shiffrin, Psychol. Rev. **91**, 1 (1984)

75. M.S. Humphreys, J.D. Bain, R. Pike, Psychol. Rev. **96**, 208 (1989)
76. R. Pike, Psychol. Rev. **91**, 281 (1984)
77. S.E. Clark, S.D. Gronlund, Psychon. Bull. Rev. **3**, 37 (1996)
78. M.S. Humphreys, R. Pike, J.D. Bain, G. Tehan, J. Math. Psychol. **33**, 36 (1989)
79. D.L. Hintzman, Psychol. Rev. **95**, 528 (1988)
80. R. Ratcliff, C.F. Sheu, S.D. Gronlund, Psychol. Rev. **99**, 518 (1992)
81. R. Ratcliff, S.E. Clark, R.M. Shiffrin, J. Exp. Psychol. Learn. Mem. Cogn. **16**, 163 (1990)
82. M. Glanzer, J.K. Adams, Mem. Cogn. **13**, 8 (1985)
83. R.M. Shiffrin, M. Steyvers, Psychon. Bull. Rev. **4**, 145 (1997)
84. J. McClelland, M. Chappell, Psychol. Rev. **105**, 724 (1998)
85. S. Dennis, M.S. Humphreys, Psychol. Rev. **108**, 452 (2001)
86. J. Arndt, E. Hirshman, J. Mem. Lang. **39**, 371 (1998)
87. A.S. Benjamin, Psychol. Rev. **117**(4), 1055 (2010)
88. M.R.P. Dougherty, C.F. Gettys, E.E. Ogden, Psychol. Rev. **106**(1), 180 (1999)
89. R.P. Thomas, M.R.P. Dougherty, A.M. Sprenger, J.I. Harbison, Psychol. Rev. **115**(1), 155 (2008)
90. P.J. Kwantes, D.J.K. Mewhort, Can. J. Exp. Psychol. **53**(4), 306 (1999)
91. P.J. Kwantes, Psychon. Bull. Rev. **12**(4), 703 (2005)
92. B.M. Turner, S. Dennis, T. Van Zandt, Psychol. Rev. **120**, 667 (2013)
93. S.E. Clark, R.M. Shiffrin, Mem. Cogn. **20**, 580 (1992)
94. J. Annis, J.G. Lenes, H.A. Westfall, A.H. Criss, K.J. Malmberg, Genetics **85**, 27 (2015)
95. S. Brown, A. Heathcote, Cogn. Psychol. **57**, 153 (2008)
96. J.I. Myung, M. Montenegro, M.A. Pitt, J. Math. Psychol. **51**, 198 (2007)
97. M. Plummer, N. Best, K. Cowles, K. Vines, R News **6**(1), 7 (2006). http://CRAN.R-project. org/doc/Rnews/
98. R Core Team, *R: A Language and Environment for Statistical Computing*. R Foundation for Statistical Computing, Vienna (2012). http://www.R-project.org. ISBN: 3-900051-07-0
99. N.A. Macmillan, C.D. Creelman, *Detection Theory: A User's Guide* (Lawrence Erlbaum Associates, Mahwah, NJ, 2005)
100. M. Usher, J.L. McClelland, Psychol. Rev. **108**, 550 (2001)
101. S. Brown, A. Heathcote, Psychol. Rev. **112**, 117 (2005)
102. M. Stone, Psychometrika **25**, 251 (1960)
103. R. Ratcliff, Psychol. Rev. **85**, 59 (1978)
104. B.U. Forstmann, G. Dutilh, S. Brown, J. Neumann, D.Y. von Cramon, K.R. Ridderinkhof, E.J. Wagenmakers, Proc. Natl. Acad. Sci. **105**, 17538 (2008)
105. B.U. Forstmann, A. Anwander, A. Schäfer, J. Neumann, S. Brown, E.J. Wagenmakers, R. Bogacz, R. Turner, Proc. Natl. Acad. Sci. **107**, 15916 (2010)
106. B.U. Forstmann, M. Tittgemeyer, E.J. Wagenmakers, J. Derrfuss, D. Imperati, S. Brown, J. Neurosci. **31**, 17242 (2011)
107. C. Donkin, S. Brown, A. Heathcote, J. Math. Psychol. **55**, 140 (2011)
108. C. Donkin, L. Averell, S. Brown, A. Heathcote, Behav. Res. Methods **41**, 1095 (2009)
109. C. Donkin, A. Heathcote, S. Brown, in *9th International Conference on Cognitive Modeling – ICCM2009*, ed. by A. Howes, D. Peebles, R. Cooper, Manchester (2009)
110. M. Montenegro, J. Myung, M. Pitt, J. Math. Psychol. **60**, 23 (2014)
111. A.H. Criss, R.M. Shiffrin, Psychol. Rev. **111**, 800 (2004)
112. M. Glanzer, J.K. Adams, G.J. Iverson, K. Kim, Psychol. Rev. **100**, 546 (1993)
113. B.M. Turner, T. Van Zandt, S.D. Brown, Psychol. Rev. **118**, 583 (2011)
114. B.M. Turner, P.B. Sederberg, J.L. McClelland, J. Math. Psychol. **72**, 191 (2016)
115. M.N. Shadlen, W.T. Newsome, J. Neurophysiol. **86**, 1916 (2001)
116. O. Ratmann, C. Andrieu, C. Wiuf, S. Richardson, Proc. Natl. Acad. Sci. USA **106**, 10576 (2009)
117. C.P. Robert, J.M. Cornuet, J.M. Marin, N. Pillai, Proc. Natl. Acad. Sci. USA **108**, 15112 (2011)

118. H. Akaike, in *Second International Symposium on Information Theory*, ed. by B.N. Petrox, F. Caski (1973), pp. 267–281
119. G. Schwarz, Ann. Stat. **6**, 461 (1978)
120. A. Tomohiro, Biometrika **94**, 443 (2007)
121. C.C. Liu, M. Aitkin, J. Math. Psychol. **52**, 362 (2008)
122. R.E. Kass, A.E. Raftery, J. Am. Stat. Assoc. **90**, 773 (1995)
123. K. Tsetsos, M. Usher, J.L. McClelland, Front. Neurosci. **5**, 1 (2011)
124. R. Bogacz, E. Brown, J. Moehlis, P. Holmes, J.D. Cohen, Psychol. Rev. **362**, 1655 (2006)
125. J. Gao, R. Tortell, J.L. McClelland, PLoS One **6**, 1 (2011)
126. D. van Ravenzwaaij, H.L.J. van der Maas, E.J. Wagenmakers, Psychol. Rev. **119**, 201 (2012)
127. R. Bogacz, M. Usher, J. Zhang, J.L. McClelland, Philos. Trans. R. Soc. B. Biol. Sci. **362**, 1655 (2007)
128. A.R. Teodorescu, M. Usher, Psychol. Rev. **120**, 1 (2013)
129. S. Roberts, H. Pashler, Psychol. Rev. **107**, 358 (2000)
130. I.J. Myung, M.A. Pitt, Psychon. Bull. Rev. **4**, 79 (1997)
131. I.J. Myung, J. Math. Psychol. **44**, 190 (2000)
132. I.J. Myung, M. Forster, M.W. Browne, J. Math. Psychol. **44**, 1 (2000)
133. B.M. Turner, B.U. Forstmann, B.U. Love, T.J. Palmeri, L. Van Maanen, J. Math. Psychol. **76**, 65 (2017)

Index

© Springer International Publishing AG 2018   125
J.J. Palestro et al., *Likelihood-Free Methods for Cognitive Science*,
Computational Approaches to Cognition and Perception,
https://doi.org/10.1007/978-3-319-72425-6